Secrets of Human DNA

What You Need To Know

By Steve Preston

1st Edition

© Copyright 2021, Steve Preston

Table of Contents

Introduction

It is a good thing you picked up this book as most people don't know what DNA is. Even if they have heard about the stuff, they don't know how it controls our lives, how it has been manipulated, or how mutations have greatly changed humans from their greatest splendor 6 thousand years ago. In this book you will find out just how amazing DNA is. We will look at how DNA affects life itself; how it is mutated; when it was mutated; how it dies; how DNA gets old; and all the rest including what makes life itself and who keeps changing it. We are going to go through the DNA mutations of humans and how man evolved or de-evolved. By the way, evolution does not simply happen as you probably were told over and over again. No matter how many times someone says it, the facts are; mutation always causes a reduction in complexity. It is against the laws of physics. We are going to investigate all this without mile-long-words, calculus, made-up biological terms, or double-talk. We will also discover things that will open your eyes to DNA strangeness including its opto-communication that can change nearby DNA so abruptly that a different species can be formed. We will also look at very promising treatments to help reconstruct or invigorate DNA While we are at it, we will destroy some of the myths concerning DNA, evolution, the Pleistocene Extinction, and even the Tower of Babel and find out answers including:

- *Why human sperm tears itself in half when entering an ovum egg, ripping away its Mitochondrial DNA. Still half-alive, the head of the sperm integrates itself with Ovum DNA.*
- *How long a DNA string is inside a Chromosome—a hint is for humans, it is surprisingly long as a DNA string is longer than you are tall and somehow, it is all twisted up and squashed*

down until every single cell of your body has this miracle of life.

- *The strange difference between alive and dead DNA. A hint is that the chemistry is all there.*
- *How DNA communicates with others nearby without leaving its nucleus home. While chemical messaging is always a factor, leaving the cell can be problematic. Miraculously DNA does something pretty spectacular*
- *Why certain types of glass disrupts DNA communication and how light affects DNA and Cancer. Knowing how Cancers regenerate using DNA helps us understand disease in general.*
- *Why ½ of all major human DNA mutations occurred at the end of the Pleistocene Epoch. There are 23 of these major mutations. They determine race, allow us to track ancestry, and can even change our species.*
- *Why the other ½ of the major DNA mutations occurred all about the same time around 5500 years ago. That's right about 12 major DNA mutations happened by the end of the Pleistocene, then nothing, the another dozen happened around 3500BC then nothing again. How odd is that?????*
- *Why the human brain size has been shrinking since 3000BC and what it means. First off, it means we are getting stupider than Neanderthal, Cro-Magnon, Boskop, Denisovan, Grimaldi and Idaltu humans and the reason is fairly well known.*
- *Why Lycaenidae butterflies have 10-times as many DNA chromosomes as humans but we believe we are more advanced*
- *Why apes have more chromosomes that humans and how apes turned into humans during the Tertiary Age*
- *How to cure cancer*
- *How Chimpanzee DNA evolved from humans instead of apes and how we know about it*
- *How one can say there are only two races of humans according to DNA. The testing has been substantial and obvious, but not usually talked about.*

- *How Neanderthal DNA is more closely linked to South Americans when Neanderthal bodies are all found in and around Europe.*
- *How DNA tracking, works and what we know from it.*
- *With a tobacco plant and a human having about the same number of Chromosomes are tobacco and humans similar? It's no wonder so many people enjoy smoking!*
- *Who the Anakim-humans were, where they came from. How was their DNA different, and how and when did they disappeared?*
- *How DNA confirms that flying transports between the Americas and Europe were common 6-thousand years ago.*
- *What DNA says about Neanderthal that helps us understand ourselves*
- *How different we are from Cro-Magnon humans of 30-thousand years ago from DNA changes*
- *What DNA tells us about the Homo-Sapien Cognatus and their cousins Gigantopithicus and Meganthropus*
- *What DNA tells us about Homo-Capensis and those similar, like Paracas, Boskop, and other groups*
- *What DNA tells us about Neanderthal and Denisovan*
- *Where various DNA defined races begin- As a hint, it was not in Africa.*
- How to cure effects of diabetes
- *Why going out in the sun causes cancer DNA mutation*
- *What STEM cells are and why killing babies to get them is absolutely stupid and evil*

First off, let me say I took 2 years of Latin in school so I could write those medical words, but I now think it is way too stupid so I will try my best to eliminate most of that stuff and speaking of the first hurdle we have the main topic Deoxyribonucleic Acid, DNA. This stuff is so fantastic and amazing, I know you are going to love learning about it. This DNA stuff is just two long strings of deoxyribose sugar with thousands of little nucleotide molecule-pairs tied to it like a ladder. Instead of staying like a ladder, it is so long, it twists around itself into a helix, then twists around little balls. Don't be afraid of DNA; they added the word 'acid' for no real reason. Putting DNA on your hair will not bleach it out or anything. This is the brain of human cells and, with 40-trillion cells, the same thing is repeated over and over again to ensure there is little or no mistakes. Unfortunately sometimes mistakes occur so we will be looking at those times as they tell us so much about who we are and where we came from.

This book is going to investigate what it is, how it works, how it dies and just about everything associated this the wonders of this incredible molecular string. Our DNA is actually split into 23 pairs of these segments called chromosomes, but if you pulled out all the DNA in a single cell it would be a ladder 2-meters long. To squash it down into a tiny the DNA strings are wrapped around little sticky balls called histones and it is these histone proteins that make the characteristic look of DNA as shown above.

I know it doesn't look like much but these things reprogram themselves, communicate by sending our photonic messages, continuously divided and replicate themselves, and are responsible for developing cancer, causing old age, mutating into all kinds of things, and so much more that not knowing about these things is really dangerous to your livelihood.

Unfortunately, some of this stuff has been hidden from us, so it's time to get back to the basics and find out how life works, how death works, how mutation happens, and how evolution to enhance our developmental state is a lie. We will find out about mutation including the strangeness concerning when DNA has mutated and some of the horrors of DNA changes in our past. To get everything to make sense we will reset your understanding of earth timing so we can track human mutations and how humans expanded across the planet; all, possibly, by looking at DNA.

Finally, we will look at how DNA modifications are now treating and curing debilitating ailments like diabetes, cancer, and so much more. While DNA is often called the cell's GENETIC material, it is so much more. Under its command, just about everything in our bodies are sluffed off and regenerated about once a month for skin and about 10 years for bones. As we age, regeneration takes longer but the main thing to understand is you are a brand new person every 10-years so don't worry too much about what you look like.

The Nucleotide molecules that form the data stored in DNA are made up of 4 proteins guanine, thymine, cytosine, and adenine, but the combinations and sequences of these 4 molecules build every bone, blood-cell, organ, skin and all the rest. While many DNA strings are hardcoded to force the development of a specific organ or trait, some DNA strands have sections that have not been programmed. We call cells with these DNA strands, Stem-cells.

The term stem cell may conjure up thoughts of some quasi-butcher-level doctor pulling apart an unborn fetus to grab stem-cells for experimentation. While that horror has happened, there is absolutely no reason for it as many cells have the "STEM" type programmability all over the body. We will be looking at some of that as we go along. Also there are some terms that you should know.

Mitochondria- This is a tiny animal that has become part of our cells. They have their own DNA. With only 40 trillion human-DNA clusters and 10 thousand- trillion mitochondria DNA clusters in our bodies some wonder if we are more mitochondria than human.

Chromosome- This is the long, long DNA molecule with part or all of the genetic material of an organism. For humans we have 46 chromosomes. Some butterflies have up to 400 chromosomes so some suggest butterflies are more advanced than us mere humans.

Histones- As I mentioned before these little sticky balls allow the DNA to wrap around and insure they don't get tangled in knots. We can call them packaging proteins if you like.

Autosome- This just means dual DNA chromosome pairs tied together. In humans all Chromosome pairs except the 2 sex chromosomes are tied in pairs.

Gene- This is just a segment of DNA that provides a specific trait like blue eyes or red hair. Classifying these things allows us to know that Neanderthal had red hair, light complexion, and a nice tenor singing voice. Never mind that stupid grunting you were told in school. Neanderthal were much smarter than present humans with a brain 10% larger than that of modern man.

Aging- The main reasons for aging include: genome damage, epigenetic-factor, shortened-telomere, mitochondria-dysfunction, and stem cell exhaustion. We will look at some of these as we go along. Here is some great news. As one study showed skin cells typically divide about 40 to 60 times in their lifetime before beginning their aging process. By genetically modifying those in order to activate telomerase production, those same cells can divide 300-times. Before you go getting treated to have young skin, there are drawback.

Telomere- speaking of getting older the telomere of a chromosome is the very end. It has not data so if there are issues during reproduction, the DNA is not affected. Over time, this telomere section gets shorter and shorter. Soon the DNA gets errors and that is not good.

Stem Cells- Like I said; you think this is talking about fetal cells that have not become an organ. Our body has many DNA cells that are still reprogrammable "STEM" cells. We will look at them later, but if you were wondering how a liver can have 1/3 of its mass cut-off and a new portion grows back it is because of the Stem type DNA strings. While lizards have full reproduction of tail and legs,   we have a few limitations, but that whole process is amazing.

Genome- the genome is the sum of the genetic information of a person or species grouped in characteristically similar sections called genes. The information is mostly stored in the cell's nucleus, and the genome is the map to build the whole organism. It contains all the information that allows our cells to build and maintain our bodies, as stored in our "human" DNA. Every fraction of DNA represents a specific characteristic of an individual. The human genome contains between 25,000 and 30,000 genes. These genes are repeated many times to make sure DNA doesn't forget to give you some eyes or whatever.

Haplotype- Geneticists can determine how many major mutations our human DNA has gone through. Instead of building a fancy name for each mutation, a simple Alpha-numeric is assigned. With this we learn so much about how when and where our DNA was modified. Haplotypes are characterized by both the Human DNA and the DNA of all the Mitochondria things we host. Rather than writing out a complex chemical equation, DNA scientists have a code. The human DNA code is written first and the Mitochondrial DNA follows like this---- $[F_m;N_f]$. This code represents and un-mutated Cro-Magnon man with un-mutated Mitochondria in his cells.

Haplogroup- Looking at when each of the haplotype mutations occurred and the sequence of each stored predecessor mutation allows Geneticists to track human dispersion and when each movement occurred over time. This capability has changed how we look at human development completely. We are going to do some of this.

Mitochondrial Tracking-By the way; Haplogrouping and Haplotyping use both human DNA and the DNA of our friendly, 40 quadrillion, symbiotic Mitochondria to help us locate and understand how we became who we are. We will look at some of this type tracking as well.

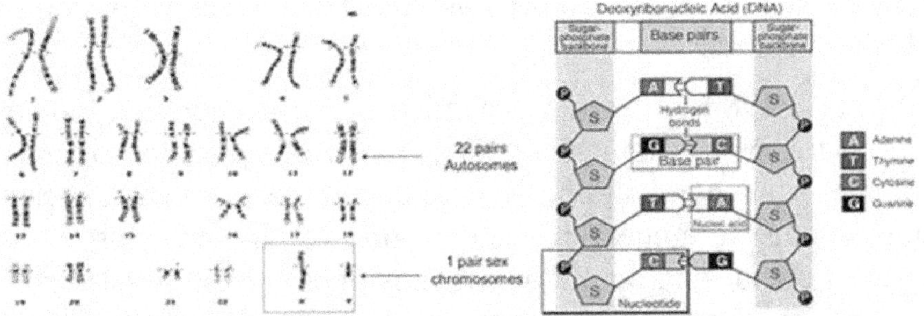

Just to give you a picture, a typical assortment of human chromosomes is shown above along with an example of how the various nucleotides build the ladder rungs of a DNA molecule. Next is an electron microscope image of 4-chromosomes and a graphic showing how those Histones shorten the DNA string so it can fit in the nucleus of a cell more easily.

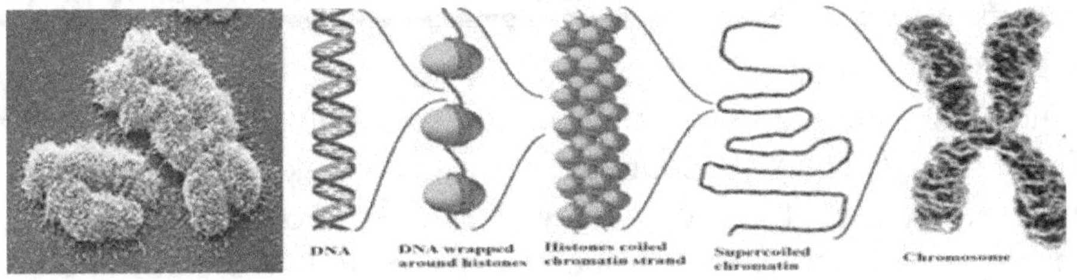

PCD Death- Programmed Cell Death [PCD] seems to be real. One might say the destiny of DNA and a cell is to die by itself. It is also referred to as "suicide by a cell." A programmed cell death is very orderly <u>and necessary</u> to destroy cells that represent a threat to the organism, such as cells infected with a virus, cells of the immune system, and cells with DNA damage. There are two mechanisms that cause PCD to occur. One is triggered by internal signals from within the cell. An example might be if a protein is damaged in some way the cell will die to insure this does not spread. The other is cell and DNA death triggered by external signals where some external damage initiates the cell to die. Cells dying are horrible to watch. First the cell begins to shrink. Then the Mitochondria breaks down making something called cytochrome-C. Then the nucleus chromatin degrades and the nucleus turns into little blobs and the DNA and RNA complex cells disintegrate.

In the early 1980s, a team of scientists demonstrated that the DNA of all living beings emit photons at a rate of up to approximately 100 units per second and per square centimeter of surface area and plants seems to generate even more. DNA controls cellular construction so sending out wrong messages can be disastrous.

If the DNA gets hurt and is not destroyed, there can be issues.

What kills DNA?-Deoxyribo-nucleic acid [DNA] is a polymer made up of three major parts; Deoxyribose-sugar, Phosphates, and Nucleotide sugars [adenine, guanine, thymine, and cytosine]

Yes, people are made out of sugar, but that is not the point here.

Nucleotides connect with each other in these long double helix structures, but some of the unions are not as strong. When Thymine and Thymine connect, it is a weak link and UV light can more easily pull it apart as shown next. Once one of these bonds has separated, the DNA can be easily mutated. We can hope that the cell just dies, but many times it tries to repair itself and we get into trouble.

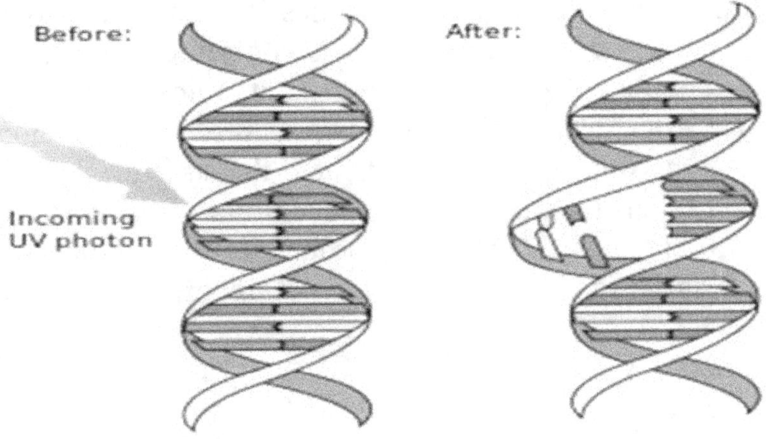

DNA absorbs photons with wavelengths of between 230 and 300nm in the UV range and it is this <u>absorption is mainly responsible for DNA damage, mutation, and death</u>. What we begin to see when out in the sun with 230nm UV hitting us is something we call melanoma as thymine-thymine bonds are broken and the DNA changes what it transmits to other cells. Optical messages ring out and mutations can occur before the DNA finally dies. There are things that can be done.

- <u>Don't let any Thiamin thiamine interfaces in DNA</u>. This might greatly limit functions in people so let's look farther.

- A second thing to do is to <u>eliminate all UV</u>. This is becoming more and more difficult. While I don't go to the beach anymore. This seems like a terrible thing to limit.

- A last thing might be to find out, quicker, when these DNA breaks occur and <u>kill off the defective strands</u>. This will take time, but it may eliminate cancers, diabetes, and brain tumors before they even start.

We will be getting into some really fun things as we go, but right now let me give you some basics you might not have known as scientists are finding we can talk to DNA using optical pulses and DNA uses this type of communication to talk to other cells.

Sometimes this is good and sometimes it is bad. We found that DNA is not some mass of sugar. It is configured as a super long, twisted, double-helix. As it vibrates, it quickly lengthens and shortens faster than you can imagine producing photons and photonic emissions it receives activated the structure to vibrate so it can tell intensity, and color of light hitting it. While DNA also reacts to chemical differences around it like other sensitive particles, this strange structure allows it to sort of talk fairly long distances.

DNA Photo-Communication

Like I said, DNA is not just another complex sugar. This thing emits, detects, and decodes visible and UV photons and this capability has fostered a brand new science that is taking off like wildfire.

Frog DNA-One of the leaders in this study of light was a Russian scientist named Pjotr Garjajev. He recently was able to intercept UV Photonic communication from a DNA molecule from one organism, a <u>frog embryo</u>, and retransmit it to another organism's DNA, a <u>salamander embryo</u>, to be exact. This caused the latter embryo to develop into a frog! OK! That would be a mean thing to do to a human, but we are just talking about a salamander. Evidently, what happens is the DNA sends out a message that can be seen by cells passing nearby, if they receive a particular optical message, they might turn into skin, or whatever. The main thing is that the old idea that *'everything was accomplished by chemical modifications making electrical differences that were interpreted by cells'* has now been greatly augmented. Think about it.

Light builds people, animals, and plants.

Let me give you an example. A skin cell senses the loss of a hair nearby. It sends out a message for the skin nearby to change to hair and a new hair pops out. If you are cut, nearby cells change to skin builders and soon, the cut disappears. This also gets into something really crazy that we will discuss later.

You are probably wondering why so much emphasis is being placed on Frog and Salamander DNA as many are getting Nobel awards and all the rest for modifying Amphibian DNA. That is because Salamanders have the crazy capability of regenerating its complete nervous system and even parts of its brain and frog DNA is so weird, many species have 4 sets of chromosomes instead of 2 and no one really knows why. Additionally it was found that Frog

DNA is closer to humans than most other animals so maybe they want to make a frog-man or something. The thing is Salamanders are filled with stem-cells so the DNA can be programmed anyway it needs to and a frog uses its eyeballs to help push food down its throat. All I'm trying to say is thousands of genetic researchers today are pushing hard to make a "better human"

Nature magazine report- A Frog genome resembles that of the mouse and the human, with large swathes of frog DNA on several chromosomes having genes arranged in the same order as in these mammals.

Science Magazine report- researchers found frog genes that are similar to 80% of human genes known to be associated with diseases.

African clawed frogs respond to human hormones so scientists are injecting female urine into them as a simple test for pregnancy. [I'm assuming if your wife brings home a "pet frog", you may soon be a happy man. No! I don't know if the frog turns blue or pink. Just leave me alone about this so we can go on.]

Xenopus laevis Frog- held the record for the sequenced animal genome with the most genes in their DNA, with nearly 50,000 genes, but the UC Berkeley team is almost finished the 150,000 Gene sequencing in conjunction with 74 scientists representing 46 institutions.

First Living Machine[1]-Scientists evolved an organism that's made of 100% frog DNA — but it isn't a frog. –These mobile organisms can move independently and collectively, can self-heal wounds and survive for weeks at a time. Some are thinking about using these animals to transport medicine inside a patient. [Yuk!]

---

1 **World's First 'Living Machine'** Created Using Frog Cells and Artificial Intelligence"-Article by Mindy Weisberger January 14, 2020

All I have to say it that with millions and millions of dollars being spent on trying to make a better frog, it won't be long before we will be making new dinosaurs and all the rest that was the catalyst for the Pleistocene Extinction. Some are trying to tell us we should not drive a gasoline car, because it will end the world when there is so much worse that can happen as we go down his DNA modification track. Maybe they can stick to fish egg and Plant DNA and stay away from making new animals which only gets our creator really mad. Getting another person mad is bad enough, but if we mess too much with our creator, he can snuff us out like a bug or just say, you are a bug and you will wake us with 6 legs. ------ Let's do this DNA changing with caution and if you believe it--- using a little prayer might even be useful as we use DNA communication to make sure large numbers of eggs hatch at the same time.

Synchronizing Fish Egg DNA

Remember I told you about making a Salamander DNA think it was a Frog, well here is another application where Egg DNA uses Bio-photonic communication between cells that are separated. This has been given a special name. It is now called the mito-genetic effect, this is the way cells can "somehow" have communications between distant samples [without fungus]. We know that bio-photons are produced at really low rates [a few dozen per second per square centimeter], but researcher Sergey Mayburov's began studying fish egg bio-photonic emission patterns. What he saw was a discernible structure in the <u>stream of photons emanating from cellular bodies.</u> Periodic bursts of biophotons resembled the patterns used to transmit binary data over noisy channels. To him, this explained how cells might be able to detect these extremely weak signals being transmitted across noisy environments.

"Technology Review" reported the following: *Biophotons from mature eggs have been shown to disrupt the growth of younger eggs. In some cases, this disruption seemed to stop the development of immature eggs entirely."*

It was found that bio-photonic radiation performs the communications between distant samples, which resulted in the <u>synchronization of egg development</u>. The photon radiation in form of short quasi-periodic bursts was observed in both fish and frog eggs; hence the communication mechanism seems to be similar to the exchange of binary encoded data via the noisy channels. It was noted that the artificial constant illumination by the visible light, over 100 times more intense, could not induce the comparable gain. The communications of some other types were reported also; for the bio-systems in the state of abrupt stress or slow destruction [apoptosis] such radiation can change the state of other bio-systems in the similar depressive way. The question he tried to answer was if the stream of photons had any discernible structure that would qualify it as a form of communication, and the answer was- yes!

Eggs to Egg Messaging-Other experiments have shown that the bio-photons from growing eggs can encourage the growth of other eggs of a similar age just like the bio-photons from mature eggs hindered the growth of younger eggs at a different stage of development.

Onion to Onion Messaging-Another Russian biologist was the first to find out plant DNA also emits, detects and decodes photonic messages. His name was Alexander Gurwitsch and he experimented on onion cells and found that stimulating one onion that was near another would cause the second one to flourish if there was quartz glass between them but NOT if silicon glass was between them. It was determined this was because the bio-photons that were being transmitted were ultra-violet that could not get through silicon glass. With no barrier or quartz, one onion being fed would cause another to react. That whole concept of talking to your plants is gone------ now you need to send the right photons and they need to be the UV light. After the onion tests, I figured hospitals were not the place to be if one was sick unless people were placed in separate rooms or something that would halt UV light separated sick people' DNA so my DNA would not get bad messages.

Fear Messaging-Plants would call out to those nearby not with pheromones [Chemical messaging] but also with light. OK! Pheromones do affect brain cells, but we are concentrating on DNA. The image following shows a cry for help when insecticide is sprayed. The photonic emission shouts out with fear and hurt and soon settles down. I'll bet you never even heard about this.

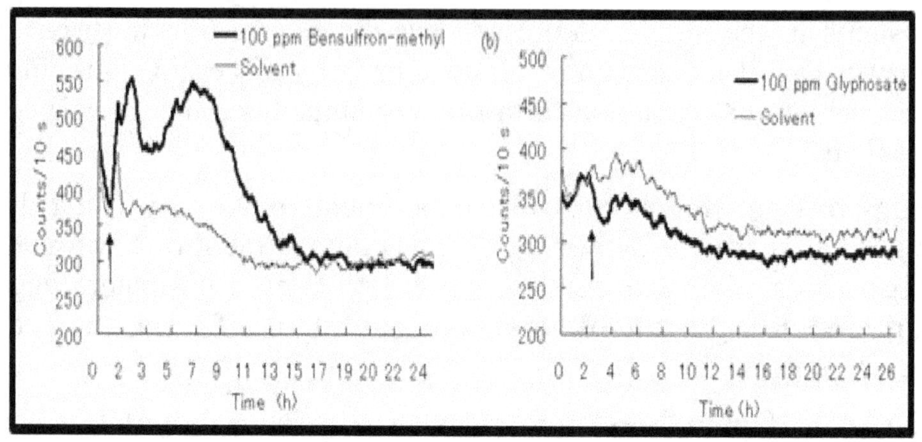

DNA Photo-repair Sensing and Reviving Itself

Here is a strange thing! It seems that scientists found that when they blasted a human cell with 380nm UV light, 99 per cent of the cell, including its DNA, was destroyed---but--that was not the end of the cell. Now for magic. All they needed to do was to reblast the cell with a very low dose of 380nm and the cell would regenerate in a single day! I know that sounds like the rapid growth of cancer so let's see what happens there.

DNA Hydrocarbons Detect and Emit

Some believe this "photo-repair messaging by DNA" may be the cure for cancer. A theoretical biophysicist at the University of Marburg in Germany, named Fritz-Albert Popp, started his work in 1970, examining differences in a carcinogenic hydrocarbon named benzoapyrene, and an almost identical but safe one named benzoapyrene. Again, UV light was the thing to activate photonic emission from the DNA of these cells, He had illuminated both molecules with ultraviolet (UV) light in an attempt to find exactly what made these two almost identical molecules so very different. The first one emitted a <u>different frequency</u> while the safe hydrocarbon reemitted the <u>same frequency</u>. He also found out that other cancer-causing hydrocarbons did the same type of photon change and they would only react to 380 nm light. Once activated,

22

a massive output of photons erupted from cancerous cells to invigorate those nearby as shown below. I know you already have a moral in your mind that you should not go near UV light if you have cancer.

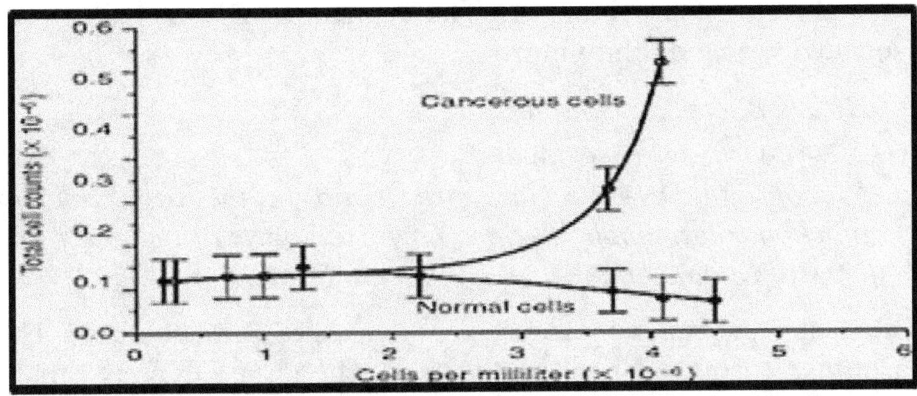

As UV comes from sunlight, there is something called Melanin, the stuff that turns tan that protects your DNA. It was determined that melanin is capable of transforming ultraviolet light energy into heat such that more than 99.9% of the absorbed UV radiation is transformed from potentially genotoxic (DNA-damaging) ultraviolet light into harmless heat, but destruction of that safeguard would quickly allow 380nm signals to begin changing cells. The next time, Melanin causes you to get sunburned instead of mutated, by very thankful. Wear those burns with honor.

DNA Generated Photons Control Everything

It seems Photonic emissions by DNA switch on and control just about all the body's processes. Given different frequencies, identical DNA in cells perform different functions, especially in the stem cells that can change functions because a section of the DNA has not been established. Sorry for that tangent, but I wanted to give you a sneak peek at some interesting aspects about talking to your DNA with light. We will look at that some more at the end of the book, but I need to introduce you to your DNA. Right now let me ask a question.

How much Human DNA is in your body of the 10.5 Quadrillion DNA clusters in your body? What percentage is human?---Absolutely correct. You get a gold star. You would not believe how many don't know that only about ½% of the DNA clusters that make us up are human. The remaining DNA clusters are associated with the following:

- *½% are Bacteria DNA clusters*
- *4 % are virus DNA segments*
- *95% of our DNA clusters are found in the little ancient mitochondria animals that are now fixed parts of our cells, but they still retained their own nonhuman DNA.*

With that in mind, why don't we look like Mitochondria? The reason is, I sort of lied in that Mitochondria DNA is small. Mitochondrial DNA is actually circular in shape, made of only 16,500 pairs of nucleotides while even the smallest human-chromosome, "21", has 48 million nucleotides. Here is another bit of information. A child will inherit mtDNA only from its mother, as mtDNA is passed down through a mother's egg. Why that is; is sort of gross. You see; millions of little DNA carrying sperm try to find an ovum egg. If they find one and they are first in line they begin entering the egg for some well-deserved mitosis, but almost instantly, the ovum shell hardens and the sperm tail is amputated. I just had a little tear for the little sperm; but try not to worry about him as his swimming days are over and he will soon transfer his partial DNA cluster to the OVUM DNA to begin a baby. His tail had all his mitochondria DNA that he didn't need anyway, but it assures geneticists that Mitochondrial DNA investigations track only the female lineage. Let's have a moment of silence for the sperm tail. The images show the first sperm entering and then the swarm unable to penetrate. Somewhere in the mix is a tail floating by of the guy in the middle.

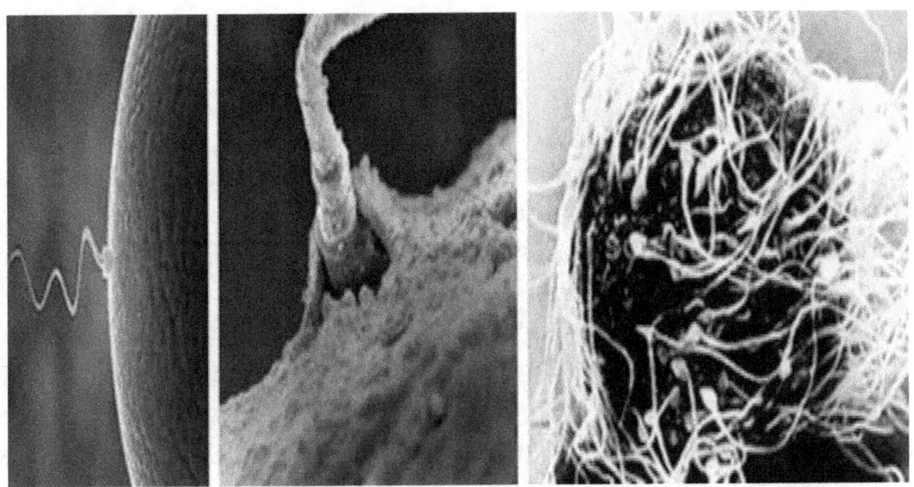

Have you ever wondered how these "I-can-find-your-ancestors" companies work their magic? Part of it is done with an alien. I don't know if you know it or not but there is a tiny alien living inside each of your cells called mitochondria. At some past time, it may have been harmful to us, but now our cells use mitochondria. For scientists they are interested in the Mitochondria's DNA as it is different than your real DNA. Mitochondrial DNA mutates differently than does our "own" DNA.

Amputated Sperm DNA

Sorry about bringing up the amputated sperm again, but because we keep losing sperm mitochondrial DNA but carry the female mitochondrial DNA we can see mutations of both DNAs and if we are testing the mitochondrial DNA, we know it comes from ONLY female ancestors while the "normal" body DNA called "nuclear DNA only traces family generations [male and/or female]. Many times the nuclear DNA is obtained from the "Y-Sex" Chromosome in a cell. As the Sperm either had an X or Y sex chromosome. It is the Sperm that always determines the sex of the newly generated fetus.

Just how big is a human-DNA cluster?-While Chromosome pair-21 has   48 million nucleotides pairs, that   is   only about 2 percent of the total human DNA in cells. The total nucleotide pairs of the 23 chromosome pairs is about 2.5 trillion. These things are so long that if you stretched out the human DNA in the 23 pairs it would be 2-meters long. Please don't try this as DNA is pretty delicate.

Speaking of that! What is life?-Let's think about what makes you-you. Everyone shouts out DNA, but what in the world does that even mean? Hopefully you will understand much more as you read the details put together for you in this book

- DNA is just a string of sugars on a protein. --- How can that bring life?
- Each person has 2 different DNA strings established in our cells the Nuclear DNA and the Mitochondria DNA.
- Cells are invaded often by Bacteria and Viruses that try to take control of the little bit of human DNA we have.
- Like most things in our universe, DNA becomes alive as it vibrates.

- Human DNA vibrates so quickly, it produces photons and that appears to be how a call can talk to another one nearby; as most of a cell is water and the wavelengths produced easily go through water. They attenuate as they go through cell walls so communication range is somewhat limited to less than 50 cells, but that is plenty to regenerate cells damages by a cut and most other damaged areas.
- Upon command, skin cells begin to get harder to establish the outside skin as other cells are sluffed off.
- Upon command, skin cells convert to new cell generators at a cut site.

Surprising number of Human Races-After testing DNA samples of just about everyone, today we know there are only 2 races of note. The details are shown next from 2 different testing methods. Later, we will see how this testing has been expanded to include Neanderthal, Denisovan and Chimpanzee DNA.

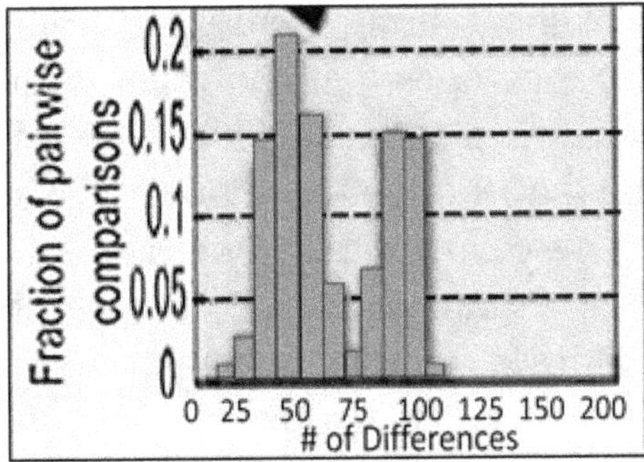

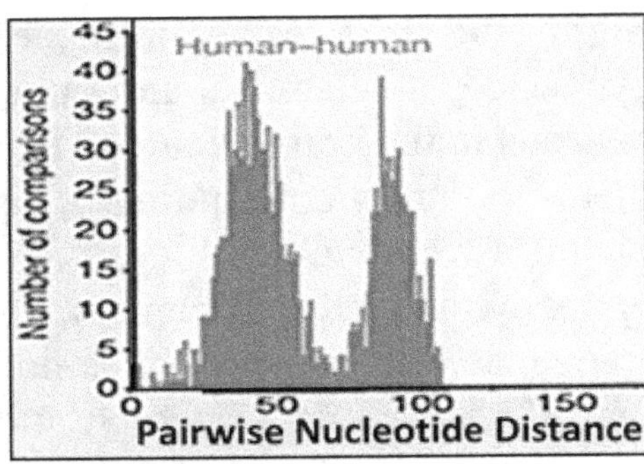

The first hump represents European human types while the second hump represent those with strong African heritage. The difference between the 2 is being reduced each year from crossbreeding. Later, we will see how this testing has been expanded to include Neanderthal, Denisovan and Chimpanzee DNA as all three are mutations of humans.

Let me make this clear. The Homo-variants alive at the end of the Pleistocene included a number of humans that did not survive the extinction event. The following shows the main groups and how they looked pretty much like we do today as their DNA was very similar, as well. All but the last one was gene spliced with the DNA of Anakim humans. I know some of you don't know about the Anakim people who dominated the Tertiary and Pleistocene periods but you will, as we go along. The main thing to know is finding DNA without Cro-Magnon similarity does not mean the strange humans like Paracas, Neanderthal, and Denisovan hybrids were "out-of-Place". It only allows us to see that early gene-splicing used the Anakim people's own DNA. We can use various DNA differences to understand something about the DNA of the Anakim people.

Sight and DNA- DNA in the cells of the vision receptors in your eye send both chemical and photonic messages that can react with neurons in the brain. Neurons take chemical-messages and convert them to an understanding-of-light that builds a picture. This includes what we call light. Light, per-se doesn't really exist as it is really just electromagnetic waves vibrating at just the right frequency to be 'perceived as light'. Other frequencies make radio waves, X-Rays, Ultraviolet and Infrared heat-sources. It's all magic, but it starts with DNA.

We could get really mystic when talking about DNA that vibrates to establish a life-force and how there is this same vibrational connection between our physical body, our soul, and our spirit, but let's just say there is a connection and go on with the physical body stuff as that can get confusing even by itself.

Ancient DNA Manipulation- While modern biologists are manipulating DNA much more than most of us know about, during an earlier time, there was much more DNA manipulation building a wide variety of animals called impure animals, unclean, or abominable animals according to many sacred works. We will have to see how all that was done, some of the results, and why it was the wrong thing to do. Besides just messing around with animals, during the Pleistocene, apes were modified, not by evolution but by DNA manipulation. The following chart shows the apparent manipulation of the animals that became human and generally when the updated animals came into being.

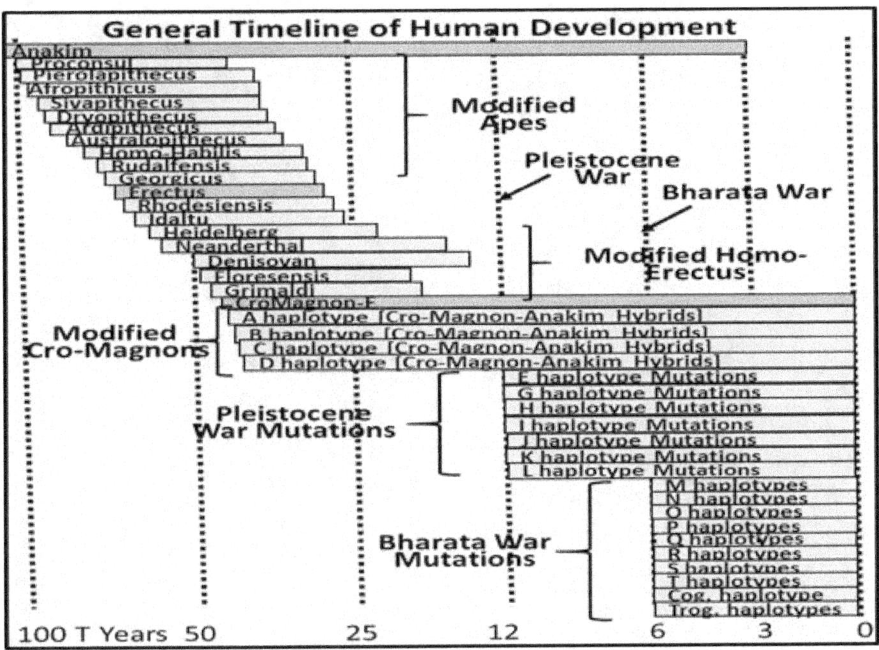

As you could see, the early Pleistocene "humans" are not exactly direct relatives which evolved into being the people we are today. Instead they are simply hybrids built up by gene-splicing and DNA-manipulations by the Anakim people of that time. Much of the time they used their own DNA for these experiments and that will help us know a little more about these ancient scientists. If we wanted to have a close relative, it would probably be the Anakim or Chimpanzee. Today we know Chimpanzee and Bonobo apes mutated from humans about 5500 years ago. Understanding this helps us understand DNA in general. Let's review a tiny portion of the testing of this important aspect of human DNA.

DNA Testing of Chimpanzees

I know this is not what you learned in school as you were told Chimpanzee and Bonobo evolved separately from man as both separated from the Gorilla lineage about 4 million years ago. Today we know that was totally wrong. Evolutionist scientists backed by consensus rather than fact will continue to lie, disregard, and weave complex reasons for ignoring evidence to hold onto their sacred self-made religion, but today we have DNA testing. Let's test some of the things we are finding out about Chimpanzee and his cousin the Bonobo to see if it makes sense chimps came from man and, most likely as part of a mass mutation that occurred 5500 years ago.

Virus Intrusion Testing-A few years ago, one group of researchers studied the genomes of 12 species of Drosophila or fruit fly, four species of nematode worm, and 10 species of primate, including humans. By comparing with other groups of species, they were able to estimate how long ago the genes were likely to have been acquired. Rather than by evolution, they determined that a number of genes, including the ABO blood group gene, were confirmed as having been acquired by vertebrates through intrusion of viruses, protists, fungi, and Bacteria. They confirmed 145 genes were acquired by this means to shape humans. <u>They found that only 50 genes were "donated" in chimpanzee</u>. This means Chimpanzee is a younger species than humans.

DNA Sequencing and Appearance Testing-DNA structure of Chimpanzee is almost a complete match with humans. While one of the chromosome strings has been split in Chimpanzee, the makeup of the DNA sugars is very close and even how the DNA clusters accumulate. The following image is of a Human, mouse, and chimp X-chromosome containing about 1,100 different genes, or sets of instructions. Each gene affects a particular trait in the body. Each specific nucleotide that makes up a DNA string [adenine (A), thymine (T), guanine (G) and cytosine (C)] show up as a slightly different shade in the image below. Notice that the Chimpanzee and Human are virtually identical. This is very strange. You can see how different even the mouse DNA is. Even the Centromere [necked down area] is the same in chimps and men.

Neanderthal and Ape Testing-The following shows the SCB1,2,3 sequences of a Modern human, 2 Neanderthal [AFxx], 2 chimpanzees [First five shown], compared to a gorilla, gibbon, baboon, rhesus, and colobus. It was so different, a portion of the gorilla sequence had to be removed to make it match somewhat better, but what we see is chimpanzee is so close it is obvious they are not descendants of the others.

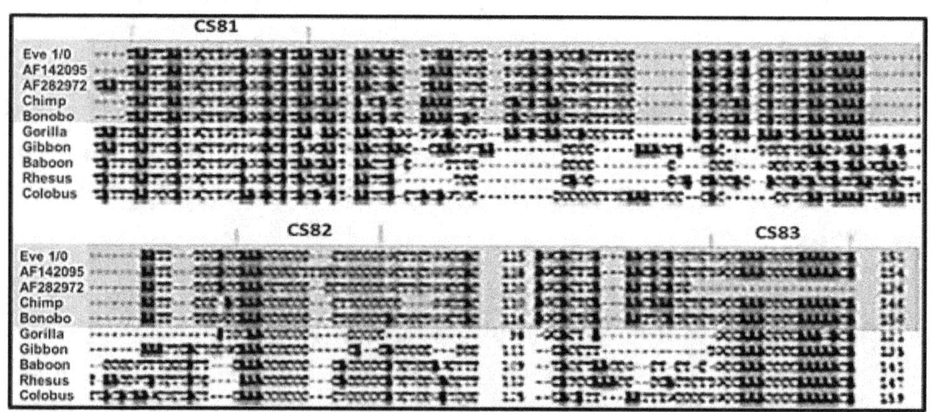

DNA Difference Testing--Human and chimp DNA was determined to be 1.2% different. Gorilla and other apes have over twice [3.1%] that much difference as they are somehow very different. In should be noted the genetic difference between individual humans today is minuscule – about 0.1%, on average. Now for the really weird part; bonobo has about 1.2% differences like chimps but Bonobo and Chimps have 1.6% difference between their DNA and Gorilla and other Apes. Chimps and Bonobo are more closely related to humans than apes.

Age of the DNA Testing- At the end of each chromosome is a string of repeating DNA sequences called a telomere. Chimpanzees have about 23,000 base pairs of DNA that are repeated. While humans only have 10,000 base pairs of DNA repeats. One could determine that Chimps have not gone through as many mutations collecting these duplicates and is a newer species.

6000-year old split Testing-It was originally believed that Bonobo was a mutated branch from Chimpanzee that occurred about a million years after the Chimpanzee and Human split 4 million years ago [using nuclear decay timing]. This has been augmented recently. In three separate studies it was determined that the human chimp split could have been only <u>6000 years ago</u> making the Bonobo split even later.

- In 2001*"Phylogenetic And Familial Estimates Of Mitochondrial Substitution Rates: Study Of Control Region Mutation In Deep-Rooting Pedigrees"*.
- In 2000, *"The Mutation Rate In the Human MtDNA Control Region."*
- In 1997, *"A High Observed Substitution Rate In The Human Mitochondrial DNA Control Region"*.

"Nature Chart" Testing-The following chart comes from the journal *"Nature"*. I wanted you to look at where they placed the "appearance" of chimpanzee up in the right hand corner around the same time as modern man. I converted the times to the newer timing but it is obvious Chimpanzee don't fit any "normal" change.

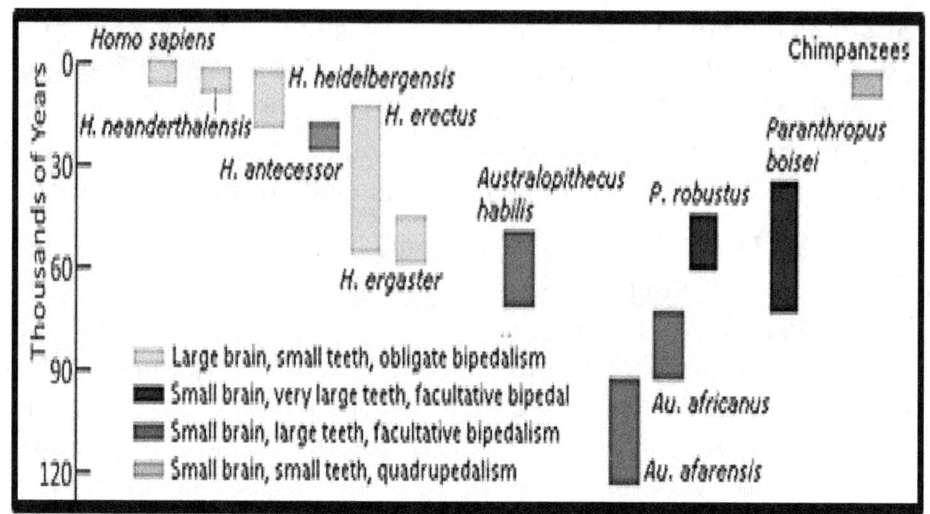

Bonobo-Chimp Difference Testing-In the study, *"Bonobo Genome Compared With The Chimpanzee And Human Genomes"* *Dr.* Eichler and his colleagues found that the human and chimp sequences differ by only 1.2 percent in terms of single-nucleotide changes to the genetic code, but <u>2.7 percent of the genetic difference between humans and chimps are duplications</u>, so we could really say it's only <u>1.1% difference</u>. They also found that more than <u>3% of the human genome is more closely related to either the bonobo or the chimpanzee genome than these are to each other</u>. They also found almost a thousand integrations of 'transposed similar sequences' [transposons] absent from the orangutan but present in <u>bonobo, chimpanzee, and human</u>. Of these, 27 are shared between the bonobo and human genomes but are absent from the chimpanzee genome, and 30 are shared between the chimpanzee and human genomes but are absent from the bonobo genome.

Skeleton Testing-The images below are of the Bonobo, Homo-Erectus, and Chimpanzee, showing what the small changes do to a person. In addition, about 25% of human genes contain "parts" that are more closely related to one of the two apes than the other.

This suggests that Bonobo did not necessarily split from Chimpanzee. It is probable it was a new mutation of Human.

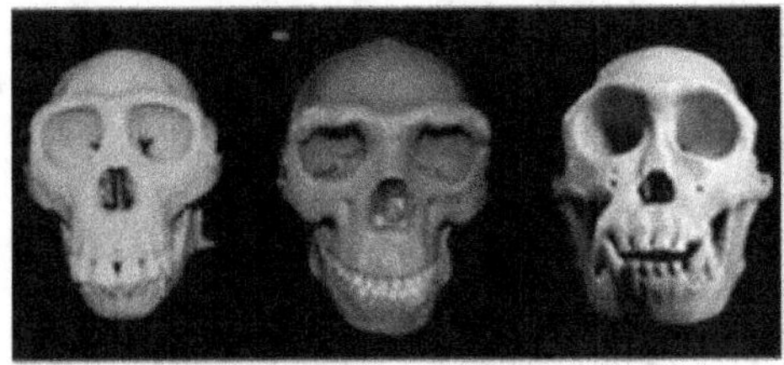 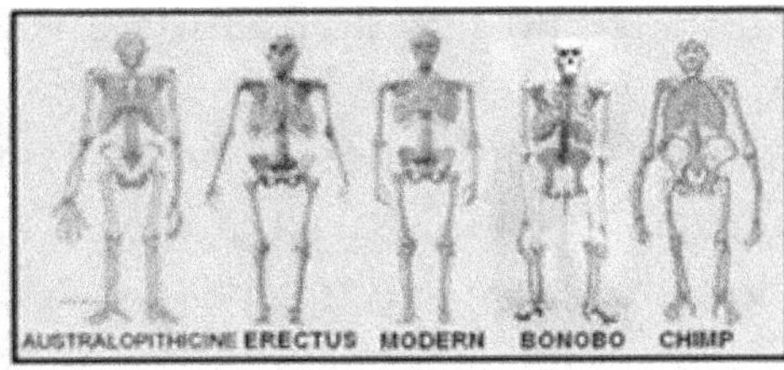

The Skeletons compare Australopithecus, Erectus, Modern man, Bonobo and Chimpanzee. While the hips and hands have reverted back to Australopithecus style, there is a lot of similarity to the bonobo and chimp to homo-erectus. The bonobo is especially similar as noted from the skull and teeth similarity shown previously. In the blink of an eye, DNA can be mutated and it has been by external events or by splicing of genes inside DNA. Humans didn't evolve from animals in Africa, so get that out of you mind.

Pleistocene Human Tracking

You have all heard about the evolution of man from a few apes in Africa who evolved into a wide assortment of ape-men living in Africa for thousands of years then they supposedly blasted out to Europe, Asia, Russia, India and the Middle East almost spontaneously leaving behind brand new type of humans because of additional uncontrolled evolution. You've seen maps, possibly heard lectures, and it is written in "accepted" anthropology textbooks. When you think about it is makes no sense whatsoever and when you look at the DNA it is even more absurd, but many are so vested in continuously trying to re-prove this crazy theory that they won't look at the obvious.

The Debunked Out-Of-Africa Model-Some 35,000 years ago, Cro-Magnons, the first "Cognizant-Observer, modern man which we use as the base for our DNA mutation tracking simply appeared one day in Europe and the Middle East without coming-out-of-Africa as was "required". Neanderthal appeared to have almost none of the DNA characteristics of the African group and Cro-Magnon seemed to just drop out of the sky. Denisovan and other late entry humans started showing this huge break between Africa and human development. That being said, we will find that a number of humans and humanoid entities in Europe did appear to have been modification of African humanoids. All of a sudden a well-spring of somewhat similar humans showed up in Europe, Middle East, China, India and other places without DNA signs of migration.

Within a relatively short time, much of the world was filled with similar humanoids as everyone wanted to leave Africa and find a new home. Haplotracking of DNA mutations showed a different view as the African hybrids seemed to stay in Africa while other humans appeared.

Haplotype Tracking-Thanks to the scientific tracking of DNA mutations we know generally where the original Homo-Erectus people lived and where they went. Generally speaking before about 50 thousand years ago, most were located in Africa with very few mutational differences. Male tracking describes 4 mutation [called A, B, C, D for shorthand description] and female DNA tracking had 4 mutation types [called L, L_1, L_2, L_3] around 50 thousand years ago the "D" haplotype may have ventured into Europe, but he would NOT seed a new group of people called Neanderthal. The early Haplotype tracking in Africa is shown below.

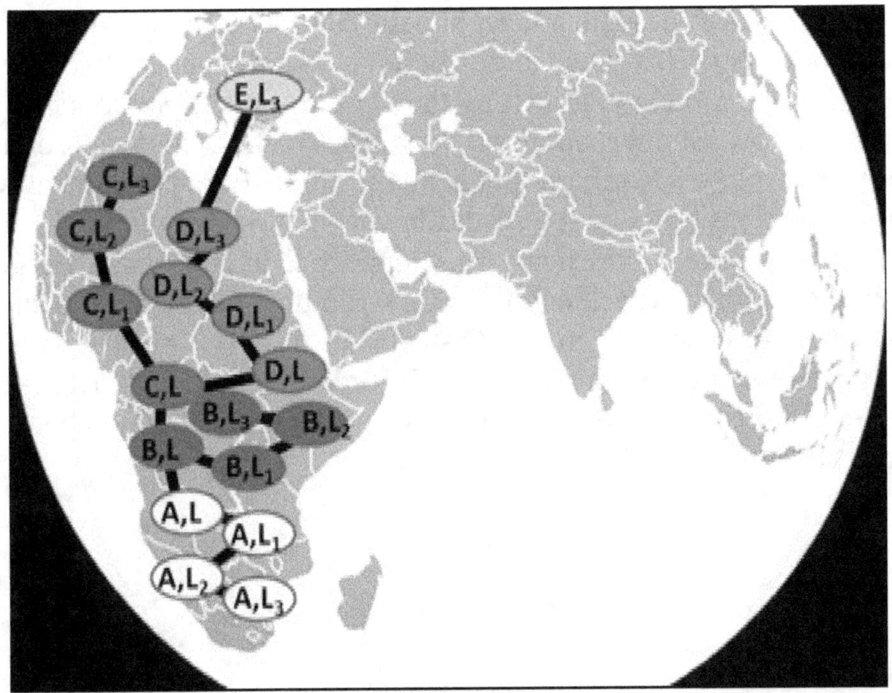

If that one isn't funny enough, the following map shows where Neanderthal type humans have been found as they supposedly walked over an even shorter period. Some would suggest these movements were done without any vehicles. Please notice these Neanderthal neve go into Africa.

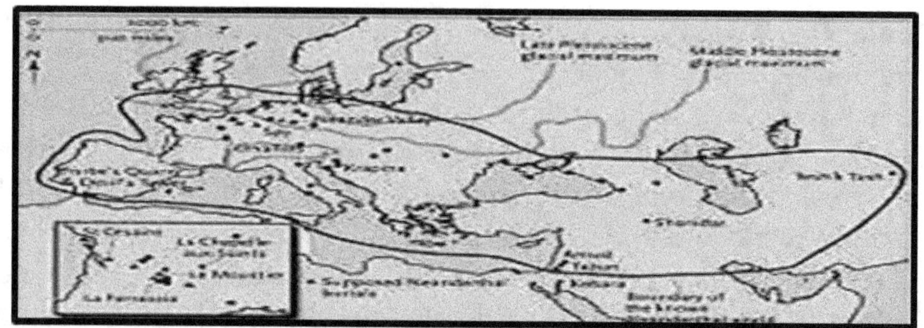

One might believe Neanderthal had evolved an aversion to Africa and a massive wanderlust to travel thousands and thousands of miles. Later, we will find the highest concentrations of Neanderthal DNA segments in human DNA is found in Australia and South America. One might believe, Neanderthals also hated Europe and got away as fast as they could, but left his bones all over Europe. Similar oddness shows up with Denisovan found only in 3 places Northern Asia, Spain, and SE Asia. Somehow, they didn't visit any of the places between. Another miniature set of humans called Floresiensis just appeared in Indonesia. Forget them being a throwback and tiny. How did they travel there to start that evolution process in the first place? Also we must wonder, why no one went back to Africa. In order for scientists to make believe humans evolved in some way they had to eliminate the Law of Entropy as if it was a lie.

DNA Evolution Misconception

I don't know if you have heard about this crazy notion that DNA magically goes against the Law-of-Entropy and, just sitting there, it can change to becoming a MORE complex entity. With this magic, animals "evolve" instead of the required de-evolution presented by ALL physicists and this Law-of-Entropy that says:

Everything WILL, given enough time, convert, or de-evolve to its lowest energy level or lowest state of complexity. It is never reversed without an outside force.

One could presumably have a person turn into a slug over time, but an animal cannot get more complex. This is why some sea shells from the Mesozoic era look exactly like they did and why insects are smaller and why jelly-fish have been de-evolved etc. I would have to say dinosaurs are not as large, but let me just stop there.

Cro-Magnon DNA were more Evolved than Modern-Man DNA

I know we have these ape-men that were changed to become less apelike and more human until, finally; the Cro-Magnon and Boskop man both had 20% larger brains than modern man and they 'just appeared'. Neanderthal brains were 10% larger than modern man's but that brain manipulation WAS NOT DONE BY EVOLUTION. Yes, we have experienced substantial de-evolved from the original Cro-Magnon with our smaller brain and reduction in capabilities, but that is also not evolution. We will look at some of that as we go along, so there is no confusion about DNA.

Evolution Problems

Speaking of inappropriate testing and determination. Evolution is a neat way to show existence of animal life that causes no fear or thought by the evaluator. The problem is that, most of the time; it doesn't work and doesn't fit the evidence as DNA will not get more complex on its own even if you wait a billion years. At best this unproven, untested determination should be called a hypothesis or a hunch, but it is taught in schools as almost a "Law". Let's look at some of the problems with evolution. Many call evolution the "No-God-Creation". After we look at the "No God creation", we will examine a number of problems with the "6-day-One-God-Creation" theory.

Too Many Animal Types-Too many animal types were spontaneously generated if an evolution theory was to be supported. We will see that after each destruction period, a huge influx of animals was apparently created <u>within an extremely short period</u> rather than the supposed "start over" that evolution theories would require. Each of the seven generally known major destructions recorded in history caused the extermination of well over 80% of all species of life and each time more animals emerged. Here is the strange part. The evolutionary life cycle didn't start over each time, but instead, many creatures spontaneously reappeared. To make things even stranger, you will see, in the next chapter, that a complete extinction record shows as many as 18 major extinction periods and many periods of minor extinctions during Earth's life cycle, which further exacerbates the problem.

Quick regeneration of life after extinction doesn't go along with evolution.

Out of Place Objects-Many things are out of place in time. Man's footprints with dinosaurs, cups found in coal, walls found deep in the Earth, and many other artifacts clearly show that manlike animals were on the Earth many years before the time that evolution can support. A sampling of these anomalies can be seen in later chapters. Don't be fooled into thinking that the artifacts are proof that the world is very young and testing methods are a sham. We will investigate some of the methods and reasons why a young Earth is not a better answer than the "evolution theory".

Out of time relics disprove evolution.

Missing Link-No crossover animals have been found. No matter how much searching, no one has found any evidence to suggest that a change in genus kind is possible- a horse has always been a horse and a man has always been a man. I know that many are saying that they have found plenty of crossover species like the partial bird partial reptile creatures [Archaeopteryx], but they truly do not represent a cross over. They really describe different separation of species that will be discussed later. If we were looking for crossover people we might look at Charles Darwin himself as referenced to Homo-Heidelberg, Darwin, Homo-Neanderthal image following. Although they are similar with Darwin having a more sloping head, it does not mean Darwin is a missing link. DNA does not operate that way!!!

41

Natural Selection Model Dilemma-Natural Selection and survival-of-the-fittest models don't work either. Shell fossils show that during some ages they get big and then small and then big again at a later time period. Horse "evolution" also shows the same characteristic. The horse started off small, got medium sized, got smaller, and then larger. All other animal fossils suggest that the adaptation to the environment does not increase the capability of an organism to survive, nor does it make a superior organism, it just kills the organism off and miraculously a new type takes its place. Sometimes it's a better animal, but <u>many times it is a "less survivable" creature.</u> *Don't worry, I will explain this stuff later.*

Types of Horses and When They Were Here-While the evolutionary time table has been artificially expanded by using nuclear decay as a standard reference, the general timelines still hold. Below describes the time table of the development of the horse using the old standard nuclear decay. Later we will look at correcting some of this, but right now, just view the strangeness of the "survival of the "whatever".

Type	*yrs. ago*	*Size*	*Toes*
Hyracotherium	55M	12 in.	Four
Orohippus	50M	14 in.	Four
Epihippus	47M	20 in.	Four
Mesohippus	40M	24 in	Three
Miohippus	35M	30 in.	Three
Archeohippus	24M	**18 in.**	Three
Merychippus	17M	40 in.	Three
Protohippus	16M	**28 in.**	Three
Hyracotherium	15M	40 in	Three
Pliohippus	14M	44 in.	One
Dinohippus	12M	**39 in.**	One
Astrohippus	10M	48 in.	One
Equus [modern]	4M	70 in.	One

God wouldn't do all this experimenting and Evolution would not allow it. Something else is most likely the answer.

Long Nose Continues our Dilemma-Sometimes "evolved features" are recognizable as mistakes. A Dinosaur mistake to be considered would be the long nosed Dinosaur. As shown in the picture his nose is over four times as long as his head and was curled back on itself. It was completely useless and there is no evidence to suggest that other dino-features evolved from this mistake. The nose couldn't be used as a battering ram like the horn extensions of other animals. It was just a long nose. For those who would suggest that this is the father of the elephant with his useful nose/hand, it would be improbable that the thing would have evolved from this characteristic to the dangly one of today. This type of mistake probably didn't come from an omnipotent creator, nor did it come from an evolutionary process. We will talk about a third option.

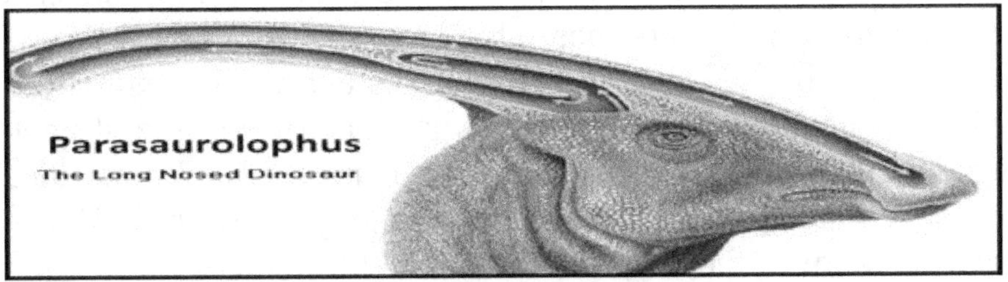

Evolution Reversal Evidence

Why are no fishes now changing into amphibians, amphibians into reptiles, reptiles into birds and mammals, and monkeys into man? If growth, development, evolution, were the rule, there would be no lower order of animals. All animals have had sufficient time to develop into the highest orders. Many have remained the same; some have deteriorated, none have evolved.

If plants and animals all developed from a one-celled animal, such as the amoeba, why did the amoeba not develop?

Animal Variety Evidence-We are told that, excluding the insect and microscopic world, there are about 3,000,000 species of plants and animals today. About 1/3 of that number are animals and about 0.05% of the animals are going extinct every thousand years. That gives us about 1 million species of animals with over

500 species being lost every thousand years. Now let's assume that the last major extinction took place about 150 thousand years ago. From that comes a very serious question.

If we start from 150 thousand years ago there would be 10 animal species doublings to approach the 1 million animal species we have today. If we assume that each doubling takes the same time period, then the last 1 million animal species would have sprung into life within the last 15 thousand years. If we further expand the 15 thousand years into 5 thousand year blocks, we find that within the last 5 thousand years there should have been over 5 thousand new species being generated if we disregard the extinction of previously generated species. We would then have to add another 25 hundred new species to make up for those which periodically become extinct [the 0.05% per thousand year figure]. I'm sure everyone knows there have not been 7.5 thousand new major mutations of DNA causing species to be generated in the last 5-thousand years. The number is very close to zero. *OK! Your uncle Jake is an exception, but you know what I mean.*

Darwin indicated: "*In spite of all the efforts of trained observers, not one change of species into another is on record.*"

Sir William Dawson; a famous Canadian Geologist said: "*No case is certainly known in human experience where any species of animal or plant has been so changed as to assume all the characteristics of a new species.*"

Not having 7500 new species in the last 5 thousand years debunks the evolution misconception.

Chromosomes- Some would suggest, as creatures evolve, increasing the genetic information contained in chromosomes enhances species. This would be easily seen as an increase in chromosome packets; or so it would seem if there were anything to evolution enhancement or uncontrolled evolution in general. Next is a short list of common animal types. Beside each animal type is the number of chromosomes used as the building instructions. Notice that "Man" is much more highly evolved than most of the animals as it has more instructions. Wow! The theory works. Man is better than other animals because it is more highly evolved.

Virus	1	Ant	2
Parasitic roundworm	2	Indian deer	6
Fruit fly	8	Mustard	10
Micro- roundworm	12	Rye	14
Guinea Pig	16	Dove	16
Corn	20	Horsetail plant	21
Opossum	22	Kidney bean	22
Redwood tree	22	Chinese deer	23
Earthworm	32	Yeast	32
Frog	36	Pig	40
Mouse	40	Wheat	42
Bat	44	**Man**	**46**
Tobacco	48	Apes	48
Sheep	54	Domestic Horse	64
Wild Horse	66	Dog	78
Chicken	78	Carp	104
Crayfish	200	Fern	500
Butterfly	380		

Ape Chromosome Count-Oh! No! It seems that we have de-evolved from the Apes by this logic. Apes, however, have one more pair of chromosomes because two sets of pairs; those called 2p and 2q, are put together in the human set as one chromosome pair, so the theory still holds as our DNA information is actually more compact in humans than in apes.

Apes and the Backward DNA-It should be noted here that almost all of the chromosomes are identical when comparing human and ape sets, so we possibly evolved from them. The only chromosome packets that differ are the 4th and 17th set. These two also are almost identical, but appear to be inverted such that the sequencing is the same but split in the middle and recombined in reverse order.

So an ape is simply an "accidentally backward human" or the reverse with a man being an "accidentally backward Ape". No one knows which came first; the man or the Ape, but evolution is still supreme.

Butterfly Masters-Now we continue down the list and find we will have to continue evolving for some time to get up to the complexity of a dog, carp, or butterfly. We should either respect our master butterflies or disregard this crazy notion of advancement by evolution. Those flying insects know that they are better than us; we don't have to acknowledge it to make it true. Chromosomes acknowledge it for us. Let me let you in on this craziness. In many butterflies and other animals, the DNA is duplicated willy-nilly. With dozens of the same chromosome, it can be bad. In humans it causes something we call mongoloid-ism and it is really bad.

Evolution isn't the Answer-I'm pretty sure you can see that the whole non-directed evolution theory has massive holes and we will reinvestigate some of these elements later, but that isn't the only theory that has insufficiently addressed the ancient timelines. Let's look at one called the Creationist Theory. -

Creationist Misconception

This is a strange on in that some believe God made the world and all the animals in six "24-hour periods", then got tired and did nothing for a day. Before I get into this one, let me first say that I strongly believe that the Bible is written in an accurate and sequentially significant way, but its time base is event driven or relation driven rather than absolute time driven. Some try to force fit an interpreted absolute time into secular dogma and when there are problems, very interesting devices are used to reestablish the "Absolute timeline" in what is now known as the "Creationist Theory". Right now let's look at problems with absolute timing without regard for other evidence as it will all make sense later.

The Creationist Theory has many problems that are pushed under the rug and the first chapter of Genesis in no way suggests the basic 6-day creation. The theory essentially is that the world can only be about 6 thousand years old, and the creation of all animals occurred within a 6-day period in accordance with the first chapter of the Biblical book known as Genesis. In order to keep this belief one must use an extremely skewed interpretation of the Biblical texts. The first chapter describes 3-'Yawms of creation' as happening before there was a sun to describe a day-night difference. The word YAWM sometimes is interpreted as 24-hours but most of the time, it means an "Age" and that is the only interpretation possible in Genesis 1. We will see that the evidence does not support this type of theory; nor does the Bible for that matter. Outside a small portion of the Christian Community, this theory is typically scoffed at, but aside from the extremely shortened time-period, many of the elements of its basis go along with much of the evidence, and so they will not be ignored in this history. Here are some of the elements that show the time base of this theory is, most likely, too short.

Magnetic Field Changes-Many Magnetic field changes have been verified by lava samples from the Atlantic Ocean and other sites. As the planet plates separate, lava pours out in abundance in the middle of the ocean, and the metallic portions of the lava align to the current magnetic field of the Earth before it hardens. Core samples from the bottom of the Atlantic show that there have been at least 170 major changes in the magnetic field alignment and none have occurred in the last 2 thousand years. A process was developed which uses these metal alignments for dating purposes. It is called Paleo-magnetic dating and it does not confirm a short time span. Each time that the magnetic field changes, havoc and destruction is rendered on the inhabitants of the Earth. By creationist standards, the 170 shifts would have occurred in 4 thousand years of improbable and unbelievable massive wobbles, destructions, and annihilations and, during all this mess, the Biblical story has almost no mention of the upheavals except for one that occurred 7 days before the worldwide flood. We will discuss in detail the destruction periods that are revealed in the Biblical texts, but they do not account for what is found in the Atlantic.

Worldwide Flood-The worldwide flood could not have created the thick pockets of coal, as implied by the creationists. Even if the flood produced wide and uncontrolled manufacture of coal deposits, the amount and depth of the deposits around the world could not possibly have been generated during that one major Earth trauma. In some areas, the coal deposits are well over a mile thick and are over an extremely wide area. The thought that the trees all congregated in one spot as they floated around in the flood and then collected to form these massive coal deposits is not probable. Trees must have been in an area, then died, grew, died again, grew, etc---for many, many thousands of years.

By the way, there was a worldwide flood when the Earth shifted on its axis during the Pleistocene Extinction 10 thousand years ago and this has been verified by a substantial amount of evidence.

Physical Similarities-The physical characteristics between the shorelines of South America and Africa, along with many other indications, strongly suggests that there was once one major continental mass that has been separating slowly over many, many years and it could not have happened over the 2000 years of creation before the flood without showing a major difference in the crustal density in the middle of the Atlantic Ocean. The crustal density is identical in the Atlantic and on the continental masses. Only under the Pacific Ocean is there any major variation in density of the crust, which brings us to another problem.

Thin Crust-The Earth's crust is thin under the Pacific Ocean. By using seismic mapping techniques it has been determined that the thickness of the crust at the bottom of the Pacific Ocean is much, much less thick in comparison with the thickness of the crust around the rest of the world. This strongly suggests that the Earth was split apart in ancient times and is slowly healing itself. By examining the amount of crustal matter that is deposited each year, the age during which the Earth was split apart has been estimated to be well over 150 thousand years ago. If God simply made the Earth and the moon six thousand years ago, why would there be the timing anomaly? I think I know what you would say. The dating methods are no good.

Carbon 14-Without a doubt we have found items that tested to be older than they actually were with this carbon 14 test method. After a volcanic action the amount of carbon 14 was decreased more dramatically than normal, which shifted the testing results but we have also found items that were dated to be younger than they actually were. This was noticed whenever the amount of Carbon 14 was increased unnaturally. Carbon 14 has a half-life of 5600 years. That type of dating is only good for about 40 thousand years and assumes that no outside force increase the percentage of carbon 14 on the tested sample. We now know that neutrinos continuously reset nuclear isotopes making things seem many times older, but there is a new timing that has fixed much of that. We will go over that briefly as well.

Oxygenated Air-It has been suggested that the air was more oxygenated long ago and therefore, the decay process was modified before the flood, more oxygen should mean more carbon-based items. This would mean that there would be a higher concentration of carbon 14 than we currently see, which would in-turn mean that any carbon-14 dating that crossed the high oxygen boundary would be in error---This error would indicate that things were actually older would test to be less old because there would be too much carbon-14 remaining. Items tested to be 6,000 years old may be twice as old, which of course is the wrong direction for the creationist belief.

Nuclear Events-An additional problem for carbon-14 dating is a nuclear event, and the Earth has seen many. This will be discussed later, but a nuclear event would increase the available carbon-14. Again there is a creationist dilemma. Increasing carbon-14 means that things are even older than normal carbon-14 testing would suggest.

Too Many Animals-There have been so many animals on the Earth that we still haven't run out of the oil that was produced by the decay of their bodies. If the animals were only here during the 2 thousand years before the flood, the Earth must have somehow been much, much, much larger to allow them to walk and not be piled on top of one another. I know that some believe that oil came from some other means, but right now let's just assume that it came from dinosaurs, because it is the most logical without additional insight.

Karoo-On the southern portion of Africa lies the greatest find of terrestrial vertebrate fossils [mostly swamp dwelling reptiles].

It is estimated there are 800 Billion; yes that's billion with a B, animals in a sandstone and shale deposit that is 20,000 feet thick.

It is stretched out for hundreds of miles. This simply could not have been a single clump of creatures pushed into one area as the floodwaters subsided.

Differences Versus Time-There are too many differences and not enough time. According to Creationist view, the flood occurred 4000 years ago and only olive skinned Adamic people survived. Within a period of about 100 years, they mutated into red skin people, white skin, yellow skin, black skin, brown skin, straight hair, curly hair, flat nose, high cheek bone, and slanted eyed variants around the world. Then, for the next 3,900 years nothing happened at all. Carvings from thousands of years ago and today show people look the same. Those first 100 years must have been something if we are to believe a flood date of 4 thousand years ago and only Noah descendants as the survivors.

More Accurate Timing

Both the Evolution timing and Creationist timing don't go along with reason or have confirming evidence, but today, scientists bored great holes in the Ice of Antarctica and Greenland to find a great method to time the massive extinction events at the end of the Triassic, Jurassic, and Cretaceous, and Pleistocene Ages. To confirm the data, they also tracked the hotspot that has been moving the Hawaiian Islands for hundreds of thousands of years and the Magnetic reversal indications locked in the magna source at the bottom of the expanding Atlantic Ocean. The following graphic shows that every 120 thousand years or so the Earth goes through massive extinction events described by massive reduction in O_{18} buried [first chart], and CO_2, buried and locked in place [3rd chart]. Changes in the magnetic field of the Earth shown in the 2nd chart adds a level of confirmation completed by tracking the direction and placement of the Equator described in the Hawaiian Islands Hotspot over time [last chart]. They all say the same thing. Very quickly we can determine both the "standard" Nuclear timing and the Creationist timing are both in error and an in-between timing is easily confirmed.

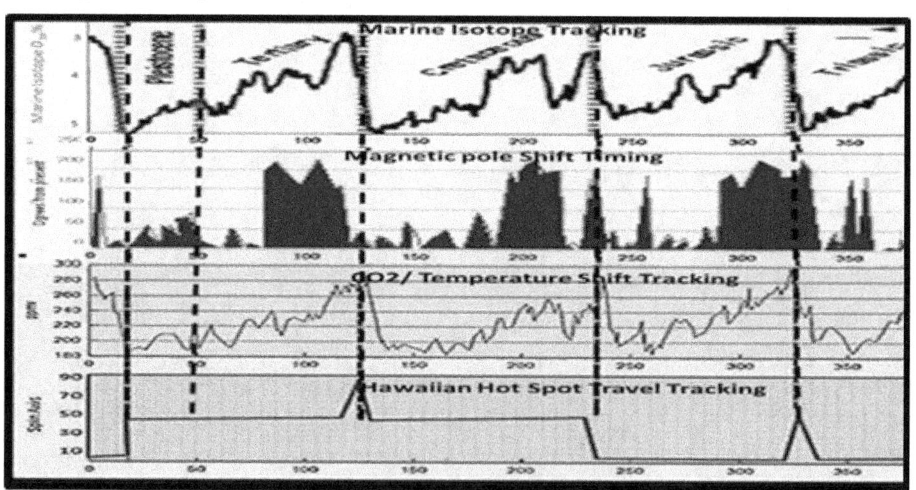

To make this even more useful, the following chart describes what appears to be a horrible worldwide war during a time called the Young Dryas between 10 and 11 thousand years ago, just before the Pleistocene extinction as shown.

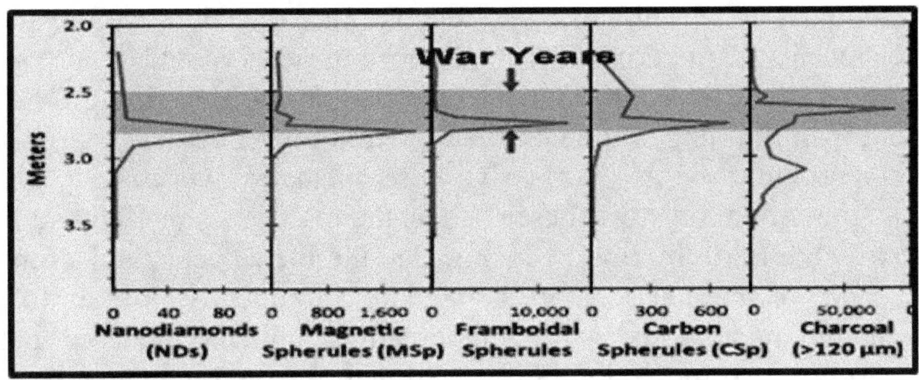

We will use these corrected timing markers as we try to understand DNA; how it changes; how it affects what we call evolution; and all the rest. So how did millions of animal types get here????????? A hint is there was outside help. This help came in a wide assortment of ways, but they all allow us to counter the Law of Entropy correctly. Several ways are listed below

- *Our creator God made new humanoid types.*
- *Previous humans had sex with primitive humans to make less primitive types.*
- *Previous humans used gene-splicing techniques using their own DNA to make better humanoids.*
- *Mutations caused by Viruses [manmade or arbitrary] changed human DNA.*
- *Mutations caused by radio-activity of some kind [man-man or arbitrary] changed DNA.*

Let's consider the Gene-splicing method.

Gene Splicing During the Pleistocene

I know some of you were never taught this, but you don't have to look long before you find that genetic modifications of animals, including human animals, were accomplished in labs before the Pleistocene Extinction. If I had more time, I could step you through all of this, but I really want to get into describing DNA more than building you a complete picture of our ancient world in this particular book. Not only were bio-labs everywhere, the changing of animals was described as the major reason God had to destroy the Earth by flood which ended the Pleistocene, and strong evidence of these actions are being found frequently. We are told in many ancient texts that mis-creating animals bothered our creator-God so much, he didn't even want his new humans to eat of the fruit-of-knowledge as it would give them inappropriate ideas. We know the story. No one listened, the Anakim people taught the newly created man [Cro-Magnon] how to "change" animals, and soon, Cro-Magnon humans were out on the laboratory floors making new animals which were considered UNCLEAN abominations by God. Besides Judeo-Christian works, we find the same issues in just about all ancient works. Here are just a few of the accounts and Yes! They had electricity and many things we had to rediscover over time. More than that, scientists tell us Cro-Magnon's brain was 20% larger than our current brains. I will describe how that all happened later, but just know we are talking about humans who were smarter than we are today.

The initial biologists were from a group of very ancient humans called the Anakim or Anunnaki, who were among those who had lived with the dinosaurs and probably "designed" many of the odd animal types. Later, a newly created 'Cro-Magnon" human would learn the secrets of genetics and become great scientists. The problem was, just like today, they really didn't know much more than splicing genes and seeing what type of animal appeared. What a mess.

Cross-referenced Details

To help confirm what we are finding out about the cross-breeding, mutation, and genesplicing of DNA in humans and animals, we find all kinds of historical references. Here are a few.

Book of Giants[2]-For the Anakim knew the secrets of heaven and sin was great in the Earth because of their experiments. They made mistakes and killed many animals and people. They selected two hundred donkeys, two hundred asses, two hundred rams of the flock, two hundred goats, and two hundred other beasts of the field. From every animal, and from every type of human was taken its seed for mixed sex.

Enoch[3] 7:5-6 and the Anakim began to <u>sin against birds</u>, and beasts, and reptiles, and fish. They have corrupted them and announce life to them.

Jubilees[4] 4:8- <u>All men, cattle, beasts, birds</u>, and everything that walks on the Earth was corrupted. -- And after this the Anakim sinned against the beasts and birds.

Jasher[5] 4:18- The sons of men took from the cattle, beasts the fowls, and taught <u>the mixture of animals</u> of one species with the other. All flesh had corrupted its ways; all men and <u>all animals</u>.

Genesis 6:4- Anakim giants were on the earth before Adam [1:24-26]-God said, "Let the Anakim of the Earth modify living

---

2 Giants- We have found 5 copies of this sacred book with the Dead Sea Scrolls and it is found in the New Covenant Bibles and Messianic Essene Bibles.

3 Enoch- This book is so sacred we have found 13 copies of the book with the Dead Sea Scrolls, it is found in the Ethiopian Bibles, New Covenant Bibles, and Messianic Essene Bibles.

4 Jubilees- This book is so sacred to the Early Christians 11 copies of the book were found in the Dead Sea Scrolls, it is found in Messianic Essene Bibles, and New Canon Bibles.

5 Jasher- [Yashar/Book of Upright] This is part of the New Canon Bible and possibly the source of Moses Apocryphon, 4 fragmented copies were found among the Dead Sea Scrolls. The book is mentioned 2 times in modern Bibles as a book of authority creatures according to their kinds. -These became unclean animals. The Anakim-gods said, "Let us modify mankind in our image. [1:27-28]-Afterwards, God created a new-man in his image. [2:7-8]- Finally, God made a new man [Adam] and placed him in the Garden of Eden

Sibylline Oracles[6]-1- *The earliest race made wars and the last calamity snatched them away from life* [Mesozoic Humans]. *-- Secondly, God formed another very subtle race and a huge body; having a haughty heart resolved to fight against heaven* [Early Anakim became angels]. *--And after these a third strong-minded race appeared. From them, many an evil and battles did continually destroy* [Some Angels rebelled against Heaven and a massive war on Earth and Heaven destroyed the Earth]. *A fourth race proceeded, who did not fear God, nor had regard for men. They caused wars, homicides, and battles.* [Some Angels were kicked out of Heaven, the Anakim had to live on Earth again] *--Yet another [5th] race much worse formed from the immortal. Since they wrought many evils, Tartartus, became their prison* [These were called the Nephadim in the Bible, but I am calling all of these people Anakim for simplicity] *-- And then, God fashioned a living product, copying a new man from his own image and bade him in ambrosial garden dwell* [Cro-Magnon with a soul, spirit, and self, in the Garden of Eden].

Book of Dzyan[7] *-The first people* [Anakim] *were self-born from the brilliant bodies of the Lords. The Second Race, Yellow skinned*

---

*6 **Sibylline Oracles**- Also called Visions of Sibyl, this text was written in Greek hexameters. The original Ocular books were destroyed in 83BC so Romans sent envoys to find copies and replaced much of the sacred work by 76BC. We are given some detail when the book first began as the original Sibyl writer was described as the daughter-in-law of Noah. The earliest transcription of the 14 remaining books was around 650AD. It is considered sacred and found in a number of modern Bibles including the Messianic Essene Bibles, and the New Canon Bibles.*

*7 **Book of Dzyan**- This ancient Tibetan work cross-compares the Genesis story* *and not born by sex* [Ape-Men]. *The third race was* <u>*evolved*</u> [Homo-Erectus]. *They were red and remained mind-less. They could not speak. The fourth race* [Cro-Magnon] *finally developed speech and were brown. They became tall with pride. They said; we are the gods. They built huge cities of rare Earths and metals* [during the Pleistocene]. *They took wives from the third race and bred* [In this book I call all of them Anakim]. *This fifth race had crooked red-hair* [Hybrid Anakim-Cro-Magnon]. *When the time came to rid the Earth of the large beasts that had been made, the first great waters came and swallowed the islands. The Guardian of Venus was surrounded by blazing masses of fire which filled the sky* [Meteors from Venus when it was destroyed] *with shooting tongues of flame. The unholy and most huge animals were destroyed. The flood continued. The holy were the only ones saved; yellow, brown black, and some red remained. Those produced from the Holy stock remained. The new land was ruled over by the first Anakim Kings.*

Book of Creation[8]*-The first man was* <u>**spirit**</u> *endowed and appeared during the first Age* [The Mesozoic Anakim people became angels]. *The second man was* <u>**soul**</u>*-endowed and appeared during the sixth Age.* [God created the Homo-Erectus man, who would be modified by the Anakim] *The third man, called Adam, is a man of the law. He appeared during the eighth Age* [Cro-Magnon] *and became numerous and produced every kind of scientific information.* [This included Genetics.]

Junius Genesis[9]*-The first created man was a noble race, the angel princes, which later perished utterly* [Mesozoic Anakim humans

---

and adds significant insight. It is not known when the book was first transcribed from Sanskrit.

8 Book of Creation- *also known as "Origin of the World" The Ebonite Christian sacred text confirms three creations of mankind, the war in heaven, the punishment of the angels defiling animals and more.*

9 Junius Genesis- *also known as Junius Codex 11 expands details of the Genesis story and was transcribed from ancient verbal texts by Fransicus* became Angels]. *For it seemed to them in their hearts that they themselves were lords of heaven. Many fell and became Anakim.* [Some were changed back to Anakim humans]*Then these Anakim took wives from the tribe of Cain.* [Hybrid Cro-Magnon-Anakim humans] *Lord resolved to punish the Anakim, and their Giant offspring. It repented God greatly that he had made man.*

***Generations of Adam*[10]** *6:1-5- Ammah, Adam's descendant, <u>understood the secrets of creation</u>. She <u>manipulated the fountain of life</u> until she had <u>created new forms of beings</u>. Tranter, her son, <u>manipulated the very nature of man and beast</u> to create new forms which God had not ordained.*

Zadspram[11] *-Satan <u>miscreated creatures</u> and they became useless. God saw the <u>defiled-unclean creatures</u>, they did not delight Him. Satan's downfall was the unrighteous creation of the creatures and ignorance.*

Let me just complain about one of the mis-created animals called the whale. Here he was with a brain 10 times the size of a man's brain and he was "made" with no fingers or arms. In fact, he was so heavy he had to stay in water just to breath and he could not breathe underwater. To make this whole thing even sadder, a tightly fitting baleen was placed in front of his mouth so he had to eat millions of tiny shrimp to survive. Not being able to vocalize words, he spends his whole life crying out in agony. God did not make the whale and a butterfly seems to be a mistake.

Massive Chromosome Butterfly

Let's get back to the pesky butterfly. Organisms like butterflies and the like that have 100+ chromosomes usually 'acquire' them

---

Junius around 700AD.
10 *Generations of Adam*- *referenced in Genesis 5, this book is part of the Messianic Essene Bibles.*
11 Zadspram – *This is the Bible of the Zoroastrian Christians. Transcription began around 700 BC. This comes from the book, Zand Akasih.*

through a process called 'polyploidy', which is a duplication of identical chromosomes. This is almost always fatal in animals. If not, it is debilitating. In humans it causes Down's syndrome so don't wish for more chromosomes just because you think it will make you into a super-man. I don't know how the Down's syndrome butterflies got these BAD chromosomes, but it does not seem to be natural. One suggestion is that Anakim were experimenting and butterflies appeared. Let me quickly describe how the Anakim got here so we can get that out of the way. As understanding them is key to understanding what we call evolution of our DNA.

Hopefully, you are beginning to believe this whole uncontrolled evolution has issues. Let's look at the first humans who lived during the Mesozoic Era with the dinosaurs briefly so we can get some more background.

Arrival of Anakim

I don't want to spend a lot of time on this section, but I do want to introduce you to the ancient Anakim. Let me say right now that I am using the Jewish name for this powerful, large brained, giant, scientifically-inclined ancestor for simplicity and because the Jewish texts have so many names for this type human; Nephilim, Nephadim, Anak, Anakim, Anakites, Rephaites, Emins, Amorites, Aviums, and so many more, that is settled for Anakim because of its similarity to names from other lands. You will see there is a lot of evidence that many of the modifications that appear to be evolution were carried out as Biological experiments. This included the modifications of Homo-Sapiens by this group called Anakim, Anunnaki, Akimim, Anu, Archaics, Araya, and other things depending on what country is telling the history.

Some have suggested the Anakim/Anunnaki came from another planet in outer-space a few million years ago, in one of the hundreds of UFOs that are seen every year over the last 3 thousand or so years. Once here, they tell us these ancient people seeded our world with space-aliens and this is why space-aliens look like us with a single nose, 2 ears, fingers, arms, 2-legs, neck, back-bone and all the rest. I don't know about all that, but there are many ancient historical records and Judeo-Christian texts which say something similar and different at the same time. Here is my overview about where these people came from and it was not from another galaxy. There are huge issues in timing with that theory even when we use Quantum-physics which eliminate time and space simultaneously with thousands of these visitors inspecting our world every year for no apparent reason.

Anyway; the first Mesozoic people, we will call Anakim and let's just say they were created by God. They lived with dinosaurs in a high-tech society built up over thousands and thousands of years. These first people were huge. We have plenty of evidence of these first people, but the things get a little weird. After 50-thousand years of advanced civilization, colonies on our nearby planetoids were common and that is why there are so many human-ish artifacts on Mars, Moon, and even on Venus are being found all the time. Most know about the 16-Mesozoic nuclear processing plants in Gibbon, Africa; the manufactured goods being found inside Geodes and coal deposits; the strange battery found inside a Mesozoic rock; and hundreds of human footprints integrated with dinosaur footprints along ancient beaches that have become solid rock today. While that does not completely rule out everything on Earth coming from another galaxy, the probability is way down in the really small numbers. Instead, we are told that, over time, most of these ancient people died and went to our linked-universe called Heaven. Let me stop here and explain that; generally, all physicists today explain our reality can't even exist without a linked universe, like heaven, to recycle energy. Just go with it in this DNA study as I don't want to fry your brain.

Near the end of the Cretaceous Era, most of the people had died and work up in the heaven-universe as no life-force or energy can be eliminated according to all the physics lovers. They were now called angels or watchers. In this book let's still call them Anakim. During the Extinction period of the Cretaceous Era, about 1/3 of the angel-Anakim rebelled to take control and they were kicked out of Heaven after a nasty war that left both Heaven and Earth in complete shambles. Moses, Isaiah, and Jeremiah all said the Earth was now "formless and void" and that "all the cities had been destroyed". Well, it seems, the Anakim were sent back to Earth where God had just made some animals. Moses indicated the "people of the earth" [Anakim] modified those animals and then these same Anakim made a "man" in their images. This was a poor attempt,   so   God 'created' what Moses called a Nu-adam [New-man] during the 6th age. Genesis chapter 6 describes the Anakim/Nephilim and many additional scriptures from around the world verify these things including "Book of Giants" "Enoch", "Jubilees", "Jasher",

"Generations of Adam", "Book of Dzyan", "Sibylline Oracles", "Age of Creation" and so many more. Most of these hand written texts were so sacred, they were copied over a dozen times, by hand and even hidden in caves for security. The most well-known copies were by the early Messianic Christians like John-the-Baptist.

God's "New-man" is believed to be the Homo-Erectus human as there was a huge change from the earlier Homo-Habilis attempts. The brain was 2 times as large, the feet and hands became man-like, and the "HUMAN" could now run on 2 legs and verbalize. The Anakim people 'integrated' with the new human and hybrids took over the land [think Antecessor, Heidelberg, etc.]. Not only was natural procreation accomplished, but many gene-splicing experiments tried to improve, God's new-man. During what the historian Moses called the '8th Age', God created still another human we call Cro-Magnon. This would have been the Adam in the Garden of Eden.

The whole time this was going on, the Anakim were making all types of weapons, scientific discoveries, and lots and lots of animals. They used this technology to 'improve' the humans by adding pieces of their own DNA. That is why were find descendants of the Anakim [specifically those found in Paracas Peru] having both Neanderthal and Denisovan DNA segments while Denisovan of West Asia and Neanderthal of East Europe are not as closely related as the Denisovan and Heidelberg humans of Spain.

Certainly, you can use the Outer-space seeding as an alternative as the same miscreation of animals, and adjoining between Anakim and Humans would stay the same; but here is the bottom line. Giant Anakim people were great scientists that lived throughout the Tertiary, Pleistocene, and first part of the Holocene. They modified humanoids to make them "better". We see the changes in DNA as genetic characteristics of a particular species or race and we have been trying to do the exact same thing. We just need a little more time.

Modern Genetic Manipulation-Today, scientists are much more responsible in what they do about making new animals. Most of us have heard about the puppies, fish, cats, and mice built to glow in the dark [1 and 2 in collage]. That is so very important if you want to find them. Others of the more responsible experiments include the fruit-fly grown to have legs coming out where antenna were [3], featherless chickens [4], big eared pigs or pigs with human organs [5], a human ear growing out the back of a mouse [6], gigantic animals like the monster bull [7], the sheep born with a human head [8], the mouse implanted with "created" memories [9], the dog with a second dog torso "bonded" to it.[10]. The eleventh is the ever popular corn that no corn-worm will eat, followed by a blue strawberry just because they can, and cloned animals. God is so very proud. The final 4 are pretty creative. The first is a donkey-zebra, followed by a pig-sheep, a Lion-Jaguar, and a lion Leopard. If you were wondering; these are not photoshopped, nor are they simple evolution of DNA. Besides this messing with trouble, we have too many genetic scientists trying to modify viruses and bacteria to build weapons like the "Coronavirus of Wuhan" weapon.

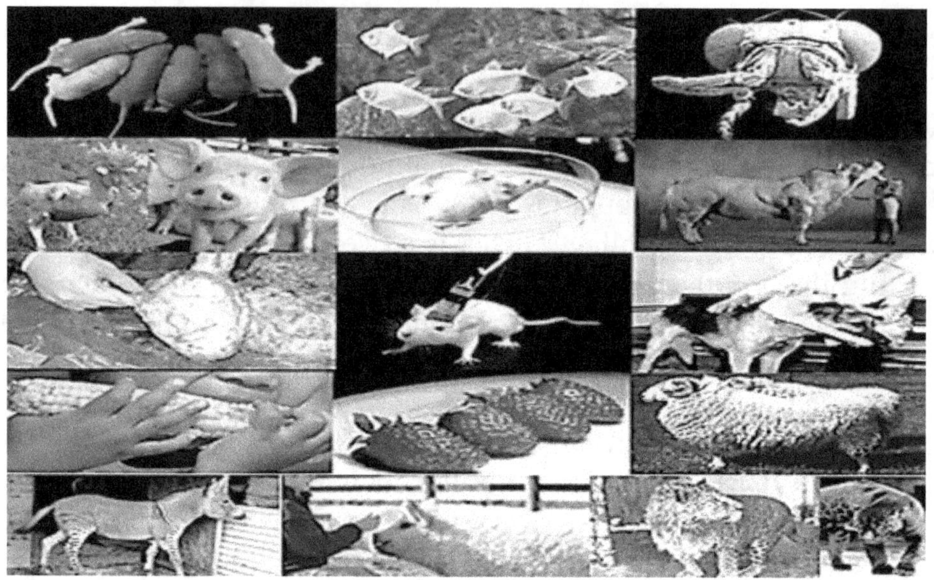

We can imagine what happened 5500 years ago; as something truly horrible began rapidly mutating DNA as it might have been some exotic cousin to the Wuhan monster, but who cares?

Secrets of DNA

Let's look at a few more secrets and get into races of people and how DNA changes everything about us.

Mutation Memory-DNA is interesting in that both the Mitochondrial and Nuclear DNA hold memories of mutation. Each time one of them mutates, the "scar" of that mutation is placed in the chromosomal string as a specific change in one of the sugars making it up. As more and more scars are built up a record of when each mutation occurred and how serious the mutation was. We will mostly look at the really serious body changing mutations to find out more about Homo-Erectus. As you can imagine there are many more major mutations in the "Nuclear" side as the record passes through more individuals, but we find that family modifications can be quantified into smaller genetic mutation so we actually have a smaller set to review.

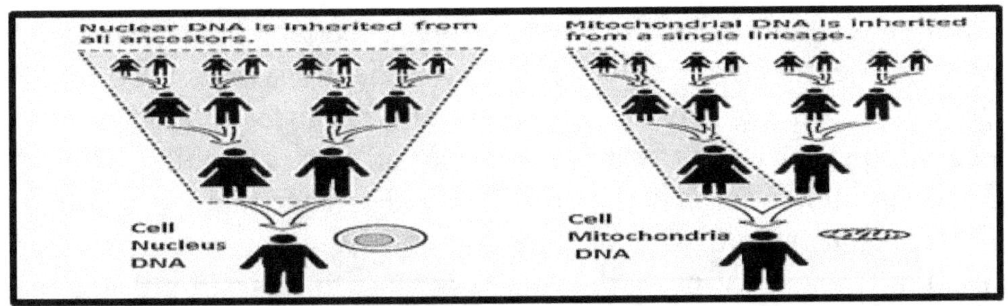

This will make sense as we go along. The most serious mutation is depicted as a capitol letter followed by a less serious one as a number and then comes a lower class letter followed by a letter. Once the coding gets too complicated, additional codes were made for "groups". Let me give you an example. [R1b] this is the Y-Chromosome mutation associated with Europeans. Depending on where on the DNA this set of mutations is located, researchers can, generally, determine when a person came in contact with a European, how many times, and the timing of each of the various mutational groups interactions associated with the test subject. By simply counting the number of [R1b] mutational groups one can tell what percentage a person is European and the Mitochondrial [mt] DNA mutations will further describe locations of encounter and percentages of various known mutational race markers. As an example the mtDNA mutation "V" indicates a Scandinavian heritage. Therefore an $[R1b_m:V_f]$ mutation signature [Y-DNA followed by mtDNA] describes a more pure European Scandinavian than an $[R1a_m:V_f]$ mutation. I know this might be confusing so all you really need to know is that geneticists can track your heritage. This whole Bio-Engineering science is generally termed Haplotyping and determination of how a group comes into being is called Haplogrouping. This is incomplete when it comes to Homo-Erectus because DNA, eventually, deteriorates. This doesn't happen when you die as your DNA after death is the same as DNA when you were alive but after 50 thousand years or so many of the DNA sugars are broken down and the history becomes incomplete.

65

As DNA doesn't lie [often] we can get a better understanding of how people got here than listening to someone spout off about all humans coming out of Africa because "Lucy" the Australopithecus Ape-woman was found near the first Homo-Habilis which was near the first Homo-Ergaster that was found. Some might, stupidly, think the rest of the world was devoid of anything more advanced that a worm, but that is simply not the way these things go.

If you noticed, I did put the caveat about DNA not [often lying] as people can still manipulate things. Such is the case of the first Haplotype family tree of major mutation in the Y-Chromosome. It is shown next. The problem is that the "CF" dual mutation never happened as F was completely separate and found in the Middle East while C, along with A, B, and D all were placed in Africa when they are really sub mutations of the Anakim who spliced pieces of their DNA to modify apes so they could make people in the first place. We also now believe, C mutation humans did not venture out of Africa, but the DNA mutation marker seems to have <u>come from Europe</u>; and there is a line to Homo-Neanderthalis very late in development time, so the whole thing is skewed just to make it look like people all came out of Africa. By the way; this chart and many like it concentrate on the major mutations. [Capitol Letter]

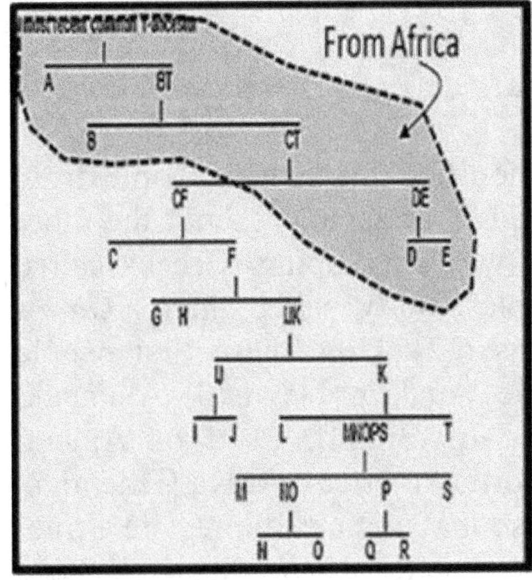

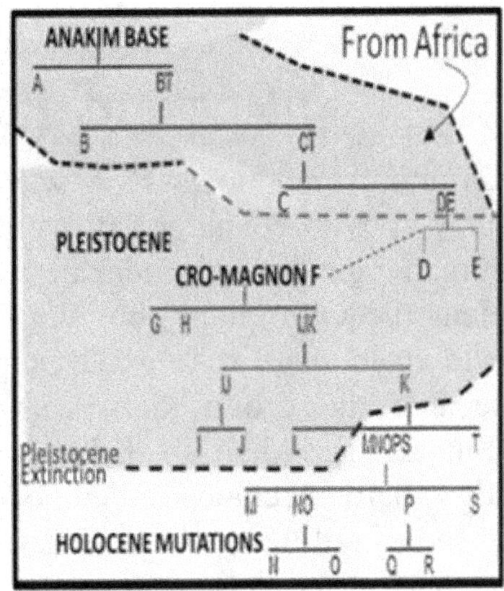

We will correct this thing as we go along so you can appreciate man's development and how Erectus and the other Tertiary human and ape-man descendants really played into modern man DNA. When we look at our DNA we find it has been changed over and over and over again---- and not to enhance our capabilities, but to make us less human. I can imagine this is disappointing, but it does make the most sense. To understand how our DNA got so messed up we need to look at Anakim, War, the shifting of the Earth's axis, and bombardment of cosmic Rays. Anakim tried to enhance us, not to help us but to help themselves. Luckily or unluckily we are going backwards from the last creation of Cro-Magnon man 35 thousand years ago. Our brain is smaller, we live shorter lives, we can't see as well, our memories a horribly dysfunctional, and many of the capabilities of our ancient cousins have been lost for all eternity.

A Race of People Use DNA

To understand how DNA was modified to change people from furry apes to Neanderthal, we need to understand about the other human race living when all this was developing. Greeks called them the Olympians, but the Bible simply called them *"Giants who lived before Adam"* [Genesis 6:2]. How these first people became the group known as the Anakim [Jewish], Anunnaki [Sumerian], Akamim [Mongulala of Brazil], Lord of Amenti [Egyptian], Archaics [Adena of North America], Araya [Dravidian of India] will be reviewed briefly so that you can appreciate a truer history, truer religion, and truer science than what has been spouted off by teachers all over the United States. That being said, just about every ancient historical record describes these people and a large amount of physical evidence has been gathered, I know some have indicated these people came from outer-space to seed our world, but that is not the most reasonable explanation. What is important here is that these people needed to modify animals to make "servants" that could do the heavy work of manufacturing, growing food, and all the rest so the Anakim could sit around and think up new stuff. A brief timeline of some of the more well-known modified Apes, Homo-Erectus, and the modification of that special man are shown next. Please notice seeming spontaneous eruption of the various human types occurring in Africa, China, Georgia, Middle East, Java, Southern Europe, Eastern Europe, and Asia without regards to the *Out-of-Africa Theory*. What it looks like to me is a massive number of genetic laboratories were in completion to see who would make the best human. I wouldn't even doubt that there was some type of Nobel Peace Prize as things got heated during the Tertiary and Pleistocene Ages. Some of the dating is changed in the chart since the radiometric timing baselines have been proven to create huge error. This in no way changes the relative timeline one can enjoy using the nuclear decay timing methods, so if you need that security blanket of timing analysis, I'm not going to try to fix that right now beyond the initial overview.

As we go along we will see some details about the DNA of this early human. We get detail by looking at the DNA of some of their descendants. Here is a kicker, the same scientists that try to push the Mesozoic Era being 150 million years ago suggest that African natives spent <u>a hundred thousand years</u> going around Africa without realizing the rest of the world was out there. We now can understand that the people, who lived in Europe for nearly 100 thousand years, are not the ancestors of modern Europeans because of a nasty flood.

Just about everyone didn't make it out of the Pleistocene Extinction. Also DNA studies of Neanderthal and Denisovan show the sub-race did not marry outside of their immediate families to spread their DNA. That brings us to what are called Large, Ancient, Dolichocephalic-Skulls of Paracas, Peru. We are told in the book of 'Numbers' of the Bible, their giant ancestors, Anakim, were so numerous in the land of Canaan, around 1500BC they scared the Jacobites to death. When Moses, the leader of the Jacobites who were leaving Egypt, sent spies to check our Canaan, they were visibly shaken and claimed the Giants were everywhere and huge. They also said the giants were devouring the inhabitants. I don't know too much about cannibalism or whatever that was, but even 4 thousand years ago, tens of thousands of these people ruled large portions of the world. This was also the case in the Americas. Just to help you remember, here is what the Jacobites saw everywhere they looked.

Numbers13:28-33 *All the people who dwell in the land are huge; we saw the Anakim there. –Those of the land devour its inhabitants, and all the people were giants; we were like grasshoppers to them.*

Let's see if we can find out more about the Anakim and how the
Neanderthal were made. The details are locked in DNA.

Paracas DNA Message

To find out about the early people who modified the DNA of so many animals, we need only look at Paracas Peru. There we find over 300 specimens of dolichocephalic skulls similar to that described as those of the Anakim. The word is Hebrew for 'giant with a long neck or head'. In Peru was found over 300 similar human remains that were different that homo-sapiens. These were not the huge skulls of the original Anakim, but the head was higher and thinner and the skull was thicker. Additionally; there was no sagittal suture interface that allows easier skull growth; the position of what is called the 'foramen-magnum' that ties to the backbone was farther back towards the rear of the skull; and hey had a very pronounced zygomatic arch (cheek bone) and the eye sockets were larger. Additionally, the Paracas people, had blonde to reddish hair that was 30% thinner than Native American hair; some of the **Paracas** skulls 60% percent heavier than conventional human skulls; and they have a cranial volume which is up to 25% larger [1700cm^3]. The first 5 shown are from Peru, but the last one is from Russia to show your these biologists were everywhere.

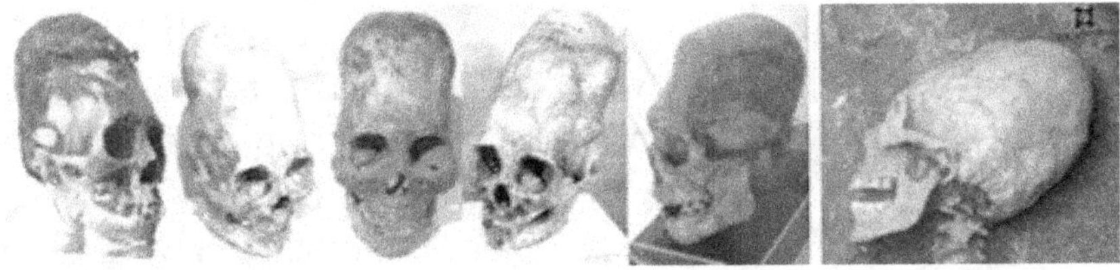

Unless you classify these people as being a predecessor human, you must face some other anomalies. *The mtDNA contains mutations unknown in any human, primate or animal known so far and there as a much higher percentage of similarity to Neanderthals and Denisovan than all other Americans.* Strangely, *from* the mitochondrial DNA in hair samples of three different individuals, they all showed a Haplogroup of H2A, Eastern Europe, closely tied to **Denisovan**. The bone powder from another skull had a Haplogroup of T2B, from Mesopotamia. Another skull showed an ancient H and H1 Haplogroup from Middle East and the Basque region. Still another infant Paracas was of the Haplogroup U2e1, whose highest concentration is in southern India. Their reddish hair and substantial amounts of Neanderthal DNA segments place them as close associates to Neanderthal. Even more important is that not having indications of mtDNA Haplogroups A, B, C or D, shows the Paracas ancestors migrated to South America DURING the Pleistocene. Like many, you might still be wondering how the Paracas people arrived in South America and how their skulls, which are NOT even Homo sapiens in shape and characteristics have the DNA of Caucasian Homo sapiens! I'll tell you.

Anakim genetically modified early humans by gene splicing or inbreeding such that this ancient human who existed BEFORE Cro-Magnon, Neanderthal, or Denisovan supplied the similar genes and not as similar genes to various experiments. With that let me start looking at some Neanderthal DNA and get rid of a couple of mis-conceptions.

Today we still find what appears to be the sequences from Neanderthal and Denisovan humans, as if some had survived the Pleistocene Extinction. Unfortunately, that, most like, did not occur. What we find mostly is Neanderthal, Denisovan, and other sequences that are "Unknown is a group of people with long thin head, red hair, just like Neanderthal, and big bodies. These giants don't appear to have been evolved from Pleistocene humans. What it shows is that the Anakim people were using their own DNA to splice into primitive ape-men to make them more and more human. Here are some more Paracas, Anakim-descendent, skulls and reconstruction of one of the Anakim people.

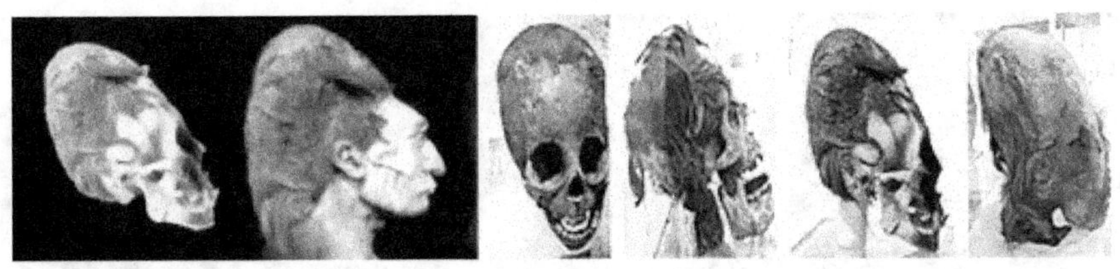

One sample had mtDNA (mitochondrial DNA) with mutations unknown in any human, primate, or animal known so far. One researcher, Marzulli Foerter, proclaimed, "I am not sure it will even fit into the known evolutionary tree," He wasn't even sure the Paracas could interbreed with Cro-Magnon hybrids. As far as physical differences, he added, "The Paracas skulls have a very pronounced zygomatic arch (cheek bone), different eye sockets and no sagittal suture, which is a connective tissue joint between the two parietal bones of the skull.

Found in the Paracas section of Peru, hundreds of the remains of these people are showing up. More are shown below including some from North America, Egypt, and another from Russia. The last 2 rows show the Paracas-like skulls of the Egyptian rulers of the 18th Dynasty from 1360 to 1290 BC. While I'm on this tangent let me tell you something interesting about the last image, Meritanten. Not only was she Queen under her dad, Akhenaten, for a couple years having a baby by him, but she became pharaoh for about a year after he died. When her brother, Tutenkomen, became Pharaoh, she was married off to King Mille of Iberia and when he died, she led the conquering assault on Ireland that would change Ireland forever. She did all that with an extremely long head.

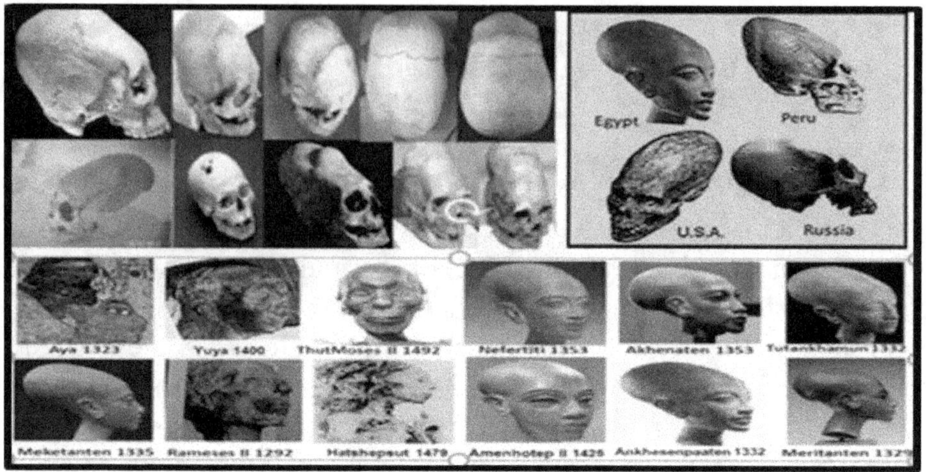

Get that whole land bridge between Asia and North America out of you head forever. These people could not have come with "Normal" people over that stupid land bridge thing. The weird part was not that this was a new race of people; it was that the DNA had "Alien" genes similar to the Alien genes found in Neanderthal and Denisovan. In this case, rather than being changed by a host, these are some of the descendants of the Anakim people. Recent DNA examinations performed on some of the hugely ancient skulls revealed shocking information that they might not have even come from known human beings. The study suggests they may have been a totally new human type. The skulls have some modified DNA which fails to match up with any identified genetic DNA material in GenBank. This GenBank is an open categorization database which contains every single piece of all recognized genetic records in the entire world. There was presented an indication that the DNA had mitochondrial DNA with various types of mutations that were totally unfamiliar in any primate, human or otherwise known to date. Some have tried to insist that the long heads were artificially made by squishing heads, but it is now known that the people were born this way. Paracas had red hair, reddish complexion, they were taller than most Americans and their long head with cranial space for a brain even larger than Cro-Magnon, tells us the rest.

As many as a hundred of similar long skulled giants have been found in North American Graves to add to the list. In SW Asia, some more were found, Australia had some, and Ancient people in Cyprus had long heads. France elongated heads were found. The Czech Republic was another place. On and on we could go. The elongated head characteristic of the Anakim was worldwide. Think of these guys as the biologists using their own DNA to make a better human. With that let's look at what we are finding about Neanderthal.

Neanderthal DNA Curiosity

Now for some real challenges. As more and more mutational data pours in every day, many are so confused they can't stand it. That is, until you add in the Anakim Biologists. Let's review what we know so far and see where it goes.

Homo-Neanderthalis DNA *has about 1% similarity to northern and East Africans and about* _0% similarity to sub-Sahara Africans._ Your guess it; Neanderthal did not come from Africa.

Homo-Neanderthalis DNA *is only about 2% similar to Asians when many remains have been found in Eastern Europe.*

Homo-Neanderthalis DNA *is much less than 1% similar to X Haplotype of the Americans. This is an important detail as Neanderthal didn't come to America, but the Neanderthal percentage to Paracas DNA is much higher. This explains a lot.*

Homo-Neanderthalis DNA *is only about 4% similar to Europeans.* This shows Neanderthal people, who lived in Europe for nearly 20,000 years, are not the ancestors of modern Europeans.

Homo-Neanderthalis DNA *is about 5% similarity to Southeast Asian populations and 8% similar to Australian aborigines.*

Homo-Neanderthalis DNA *is only about 1.7% similar to Denisovan DNA*

Denisovan DNA *is 3% similar to Southeast Asians, which is less than Neanderthal.*

Homo-Neanderthalis DNA *is more closely associated with Heidelberg man than Denisovan. This just means Denisovan was a different experiment by the Anakim not associated with Heidelberg.*

DNA scientists limit their studies to the 10% of our DNA responsible for building proteins. Of the 90% of DNA considered as "junk". Scientists have found 145 'alien' genes not traceable back to any of the DNA definable ancestors.

Neanderthal %-For the average person about 4% of your DNA is known to be from Neanderthal;

Melanesia-5% of the genome of Melanesians DNA came from Denisovan.

Tibetans- *These people have a DNA haplotype mutation that assists with adaptation to low oxygen levels at high altitude and <u>Denisovan had the same mutation</u> showing their close relation.*

Europeans-*About 2% of the DNA of all people with European ancestry can be traced to Neanderthals and a higher percentage is found in Asians, including 6% for Australians a large percentage of the Paracas of South America.*

Brazilians- *Brazilian and Native Americans have a relatively high genetic contribution from the Denisovan/ Neanderthal.*

Similar Alien genes *have been found in Neanderthal and Paracas, Anakim-like, skulls.* Sounds like Neanderthal got to South America without actually going there or they came and were confined to the Anakim/Paracas group.

Protein Similarity-*Only 87 genes responsible for making proteins in cells are different between modern humans and Neanderthals. Some of the differences involve ones involved the development of brain cells.*

Anakim DNA-*The Denisovan [variant of Neanderthal] share up to 8 percent of their genome with a "super archaic" [unknown species] that dates back to the Tertiary Period.*

The 8 percent Anakim DNA tells us the Anakim had something to do with the creation and development of Neanderthal. They simply were not evolutionary mutations from Homo-Erectus. Anakim made them. To understand this change a little better we need only read a small section in some of the ancient texts to expand on what I already presented earlier.

Enoch 2:18- *Anakim came down and copulated with women and had offspring.*

Nag Hammadi Creation Text-*The Anakim gods said let us cast our seed into her [Cro-Magnon female], so whatever she will be under our charge.*

Genesis 3:4-16- *She [Cro-Magnon woman] said, Nahash seduced me- And God put enmity between his seed [DNA] and her seed [DNA].* Before this happened, we are told, Eve was pregnant with Cain and his offspring could mate with Anakim.

Melchizedek-*Pray for the offspring of the angels and the men that came out of their seed. They were not the true Adam* [They were hybrid Cro-Magnon and Anakim]

Greek Mythology --*Zeus [one of the Anakim] had a woman created from clay [Pandora]. She opened the box of knowledge and released evil into the world. Latter Zeus had children by normal humans.*

A general statement must be considered as this is not showing a discontinuity of Neanderthal travels as it would seem. Instead, it shows Anakim DNA was used in the production of Neanderthal. Denisovan and other humans were separate "experiments. Sure enough, the remains similar to Anakim show Neanderthal traits not because Neanderthal was an ancestor of the Paracas, but an "off-spring" by way of gene spicing. Neanderthal didn't evolve from some African ape-man. He was manufactured and later traveled to Africa. Neanderthal did not survive the Worldwide Flood, but those who made him did survive. The 8% Australian similarity simply means, the same group of Anakim probably gene spliced the Aborigines. Let me show you the ACCEPTED evolution chart I got out of the Encyclopedia.

Curious Mutation Charts

I know some are thinking this is way too strange, so I thought a couple of charts would help you. First off, there is no question that evolution occurs but in order for modifications to not push a species further and further into disorder [Law of Entropy], some outsider MUST modify the chain of events to assure advancement of a species rather than the slow destruction of entropy. From the text I just presented there is little doubt the Anakim people had violated the laws of God by "miscreating". Some of the miscreated were considered human. Many of the miscreating experiments were just biologists having fun; completely unconcerned about the outcome. If you look in Encyclopedia Britannica, they present the first chart shown next as the development of animals and when they occurred. Essentially, they stated that for millions of years there was about 500 different families of marine animals then in the blink of an eye, so to speak, there were 1900 families. I modified the chart using the new dating system and it still looks impossible without biologists making new animals every week or so but it is not Twilight Zone weird.

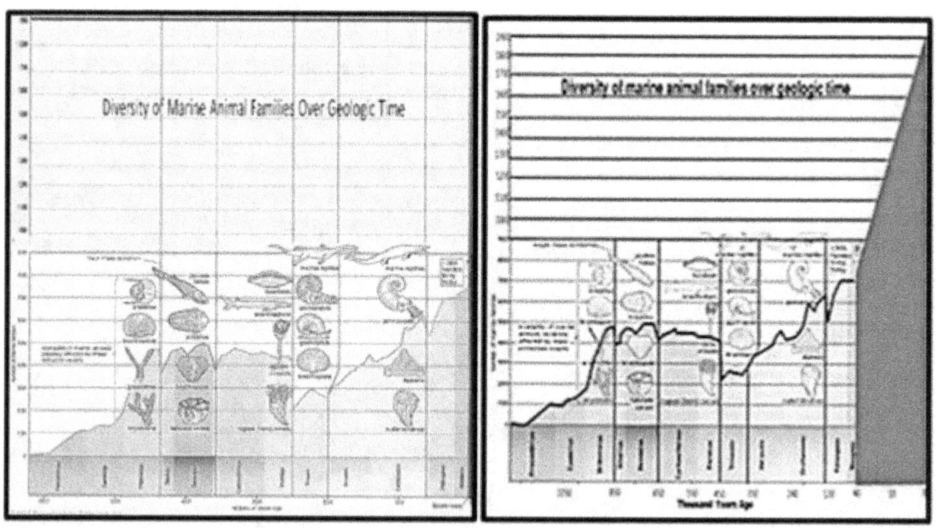

From this chart and just about any chart you can come up with, there can be little question about the ANAKIM miscreating all types of animals <u>including the remanufacture of dinosaurs</u> and the manipulation of apes and humans.

Human Mutation Chart

While we are on strange charts let me briefly look at DNA mutations. The chart following is the general details of major mutations of the Y-Chromosomes of humans. DNA Haplotype scientists record the combination of these mutations and when each one occurred by looking at the DNA. In this case, the "Y-Chromosome" track of DNA in the nucleus of each cell tells them your family history back to the first "human". For most people this track goes back 40 thousand years, but for individuals with a strong black background, this sequencing goes back an estimated 100 thousand years. While the first 2 mutations are during ancient times, The F indication appears not to be a mutation, but is the base Cro-Magnon indicator, C, D, E, G, H, I and J on the other hand show that someone was messing with the DNA of Cro-Magnon during the very last years of the Pleistocene.

<u>Homo- Erectus</u>
A="Y-DNA Adam" [100 thousand years ago]
B= Sapien [50 thousand years ago]
<u>Cro-Magnon</u>
F= Adamic [40 thousand years ago]
C= Negriod [20 thousand years ago]
D= West African [12 thousand years ago]
E= Nubian [12 thousand years ago]
G= Armenian [12 thousand years ago]
K= Japheth/Asian [12 thousand years ago]
H= Afghan [12 thousand years ago]
I= PreGreek [10 thousand years ago]
J= Hamite [10 thousand years ago]
T= Near East [6 thousand years ago]
R= Scythian [6 thousand years ago]
M= Russian [6 thousand years ago]
O= Oriental [6 thousand years ago]
P= East Europe [6 thousand years ago]
Q= American [5 thousand years ago]
S= SE Asia [5 thousand years ago]
L= Dravidian [5 thousand years ago]
N= Scandinavian [5 thousand years ago]

We will look at that chart again later. The following chart shows the same major mutations of humans by quantity since humans DNA mutation found in humans. Of course the first 2 were simply experiments with Human DNA being seeded with early hominids. Anyway, do you notice anything odd?

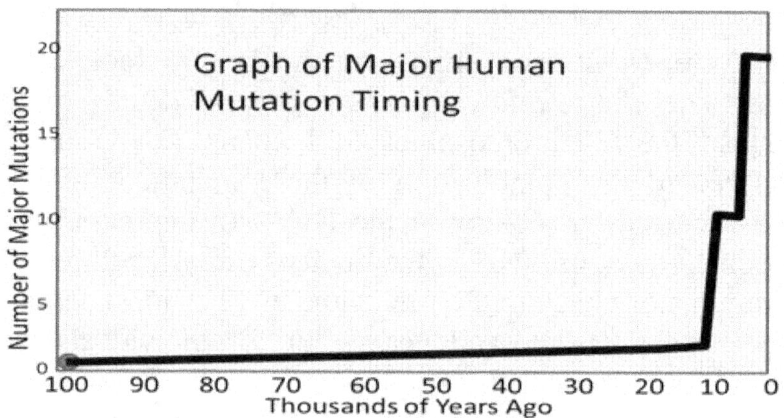

This Is Not The Chart Of Random Mutational Events

This shows that humans did not mutate at all for almost 100 thousand years, then something happened, and then something happened again to change humans. The changes can most easily be understood if we assume they were helped by Anakim genetics.

One reason it is hard for some to accept the thousands of bits of evidence concerning the Anakim people and their influence on the development of huamns is that there aren't any more living so their existance can easily be swept under the rug.

Over 86 percent *of the harmful single nucleotide mutations arose between 5 and 11 thousand years ago.*

Oddly, since *then there have been no single nucleotide mutations.*

From DNA mutation analysis *researchers now indicated that about 81% of the single-nucleotide variants in the European sampled and ONLY 58% in the African DNA sampled arose in the past 5,000 years. This shows Africans have a much older DNA* *than non-Africans and the 2 groups had very little crossbreeding until well after the 11 thousand year old massive mutation.*

Cro-Magnon appears *to have first been in the Middle-East and Migrated to Europe and to Asia well before their hybrids ventured into Africa.*

Why So Much Interest In Africa?

Earth tilt data suggests the reason Homo-Erectus covered almost all of Africa and Southeast Asia was that they were all on temperate zones. [See following graphic of present, right image, and Pleistocene Earth tilt, left].

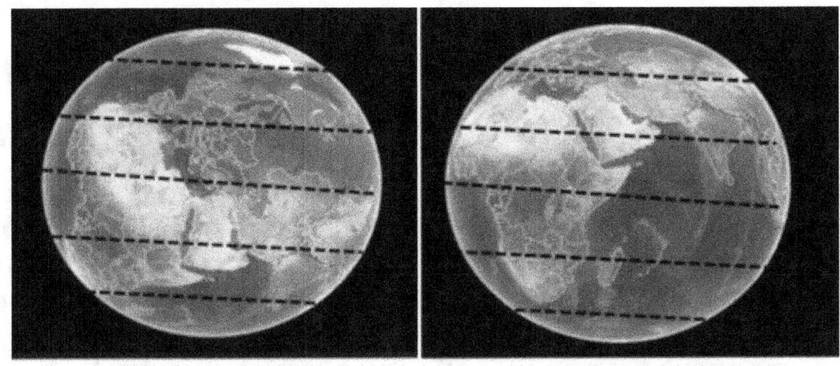

Pleistocene Axis   **Holocene Axis**

DNA tells us the following:

Major "Races" Simply Appeared- Homo-Erectus, Neanderthal, and Denisovan humans simply appeared in different locations. As one can determine from the initial details already presented there is very little probability that these substantially different individuals evolved without using the previous human DNA unless some outside force manipulated the entire DNA strings all at once. In this way 98% of the Denisovan DNA changed from Neanderthal and 98% of the Cro-Magnon DNA could be changed from Neanderthal as well and also have almost no similarity in the Denisovan DNA. To make this even stranger, when this happened, there were enough replicated Denisovan and Cro-Magnon beings to all for procreation. While we do not have enough Erectus DNA

83

to support reasonable DNA studies, we can tell that whatever happened between Habilis and Erectus was huge as there are so many physical differences we can see Habilis was another Ape and Erectus had become a man. As no viable DNA has been taken from the ancient Homo-Erectus some of this is going to have to come down to guesswork. Before getting into Erectus, let me very briefly explain the new timing standards for geologic time since nuclear decay has turned out to be so very unreliable and almost always presenting false readings that are much more ancient than known date.

Cro-Magnon Man Just Appeared-Finally, Cro-Magnon Man or Adamic man was created during the 8th Age after the Earth became void [according to Genesis 2] which we typically call the beginning of the Pleistocene Age, 35 thousand years ago. As the Pleistocene Age ended, 10 thousand years ago, massive mutations caused substantial change in humans and many racial traits were formed. We know this happened by detailed Haplotype DNA mutation tracking which has changed the way we view history. The Mesozoic Anakim Giants started it all. Homo-Erectus began changing, mostly by outside forces and then Cro-Magnon came along. The Homo-Erectus people stayed more to themselves but after years they would finally integrate and be modified to form a number of different types of Erectus based humans. After Cro-Magnon came along they mixed, either sexually or by test-tube, with the various types of individuals still extant. If you remember the previous chart on DNA differences [shown again –next], no one has really done an in depth review of these <u>two types of modern humans.</u>

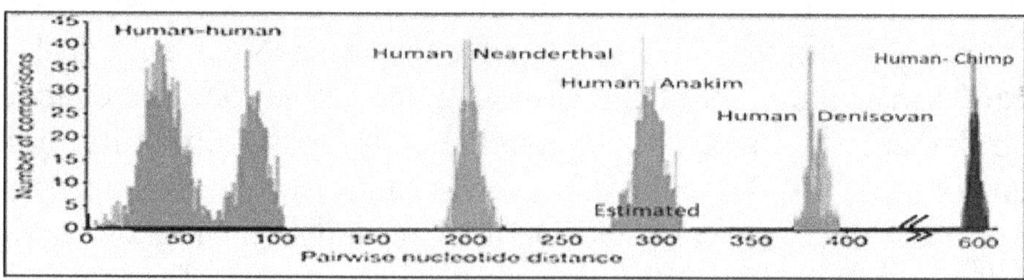

The following table can be used to provide you with an understanding about the evolution or controlled-modification of humans including the Anakim and Chimpanzee races.

Human Characteristics	Anakim	Ouranopithicus	Homo-Habilis	Homo-Erectus	Gigantopithicus	Heidelberg	Neanderthal	Floresiensis	Cro-Magnon	Modern man	Chimpanzee
Age [TYA-old]	200	100	90	80	75	70	65	60	40	6T	6T
Brain [cc x 100]	20	3	4	8	8	13	14	11	15	13	4
Height [In x 10]	19	3.6	4.8	5.2	15	6.9	6.9	4.8	7.2	7.2	6.5
Communal Life	Yes	No	No	No	Yes	Yes	Yes	Yes	Yes	Yes	Yes
Religious	Yes	No	No	No	No	Yes	Yes	No	Yes	Yes	Yes
Cranial Ridge	No	Yes	Yes	Yes	Yes	Yes	No	No	No	No	Yes
Speak and Sing	Yes	No	No	No	No	Yes	Yes	Yes	Yes	Yes	No
Man feet/hands	Yes	No	No	Yes	No	Yes	Yes	Yes	Yes	Yes	No
Run on 2 legs	Yes	No	No	Yes	Yes	Yes	Yes	Yes	Yes	Yes	Yes
Complexion	Light	Dark	Dark	Dark	Dark	Light	Light	Light	Light	Both	Dark
Human canines	Yes	No	No	No	No	Yes	Yes	Yes	Yes	Yes	No
Tool-making	Yes	No	No	Yes	No	Yes	Yes	Yes	Yes	Yes	Yes
Anakim DNA	Yes	No	Yes	No	Yes	Yes	Yes	Yes	No	Yes	Yes
Location	Many	Trky	Africa	Asia	China	Euro	Euro	Java	Arab	Many	Africa

All these changes and so much more was controlled by DNA. Sometimes the DNA had outside help and gene-splicing techniques must have been amazing back in the old days. Sometimes modifications were accomplished by external bombardments of UV, neutrinos, and gamma radiation. Sometimes DNA did change by mistakes in the DNA chain itself, but most of the time, DNA correctors inspect inappropriate combinations of nucleotides and fix them to keep us who we are. Whatever changes DNA or mutates it is registered in the DNA strings themselves so scientists later track the mutations, heritage, flow or races, and all sorts of things. Specific Mutational groups are called Haplotypes and tracking the modifications by generation is called Haplogrouping using both we get a pretty interesting picture.

Haplotyping DNA

With this new perspective and a shortened timeline let's look at DNA in more detail. DNA is a secret part of you that tells where you came from. Before we can see the definition of Haplotyping, we must understand its evolution. This starts with something called a haplotype. Simply put, a haplotype is a colony of individuals that have mutated the same way. It seems that major changes in characterizations are sort of recorded in the DNA structure such that one can look at a sequence of parts called Alleles that are stuck onto what can be referred to as a "Genetic Segment" and determine your ancestral "Haplotype". People with similar Haplotypes are similar because their DNA is somewhat similar. This Haplotype thing not only clusters people by characterization but also when the "grouping" was formed so one can determine differences of people groups, societies, continents and how these groups moved from one place to another. I know you have seen DNA used in a courtroom to tell is someone killed another if the killer left a hair in the room, but I'm talking about using this stuff the trace humans back through time to their origins.

Ancestry-Two different Haplotypes can be used to determine ancestry. One is found in 'from a male" Y-chromosome [Y-DNA] and the other is from mitochondrial DNA [MtDNA]. These Haplotypes have different designations. Haplotypes pertain to deep ancestral origins dating back many thousands of years. According to "some" research, Y-DNA is passed solely from father to son, while MtDNA is passed down the maternal line to both sexes.

Mitochondria DNA-This whole mitochondria thing should be getting you confused. If not, let me enlighten and confuse. When someone talks about Mitochondrial DNA, he is NOT talking about the DNA that makes you who you are. Instead, people have a second set of DNA inside themselves from aliens called mitochondria. Mitochondria are small "quasi-animals" that lie in the cytoplasm of your cells. It is believed these were once completely separate bacteria, but over time, they became "part" of the cells and help supply energy to the nucleus [where the 23 pairs of "real" Chromosomes are hiding as shown below].

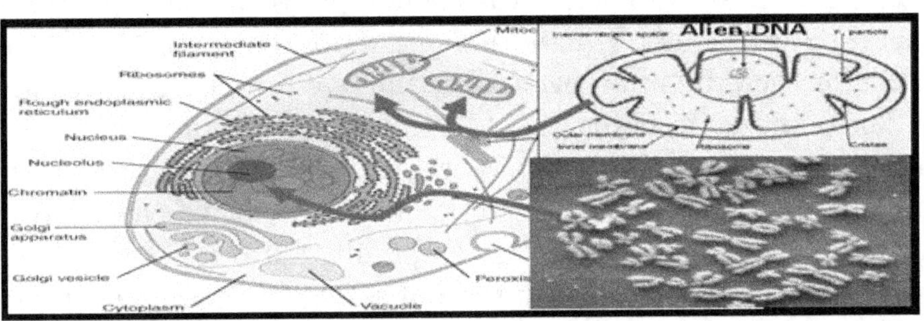

To show it is an alien, the mitochondrion has its own DNA different than that of the cell that is copied from Mitochondria in the mother [most of the time]. If the mother had gone through 50 mutations since the animal or person became who she was when she carried a new baby, the mitochondrial DNA that was passed on would be an exact duplicate and carry the information of those 50 changes or mutations. All one must do is decode those 50 codes and every change ancestor and mutation would be known. How simple is that??? To make it easier, some scientist calculated that there was supposed to be 7.5 mutations every million years. Unfortunately, for those trying to make this simple; mutations are very randomized and clumped together during various cosmic or high radiation events. A number of estimates have the mutation rate at MUCH higher rates making all the timing much more compressed than previously determined. This is where that new timing chart will be handy.

Haplotype Tracking

As I said, by testing large groups, one can map out where individual groups came from, and how the lineage took control of various places around the globe. The following maps show a generalization of this type of Haplotype flow-map. These are general descriptions of the various mutation grouping and a time-period for each mutation event. The relative timing of the events is similar to known tracking, but I have compresses the timing so we can be closer to the ballpark. While we cannot get an exact time, we can determine what mutation comes first so we can adapt the sequence to known events. Hopefully, from my previous discussions it is known that this is not an exact science. There have been so many intermarriages and so many thousands of years, who came first, the Hamite or the Gaelic, can make it difficult to track entities and attempt the generalization of how a person got to be who he is today. Also, note that I confined most of the African mutations to Africa. While there certainly were extrusions, especially by the "E" grouping, these would not happen for a while.

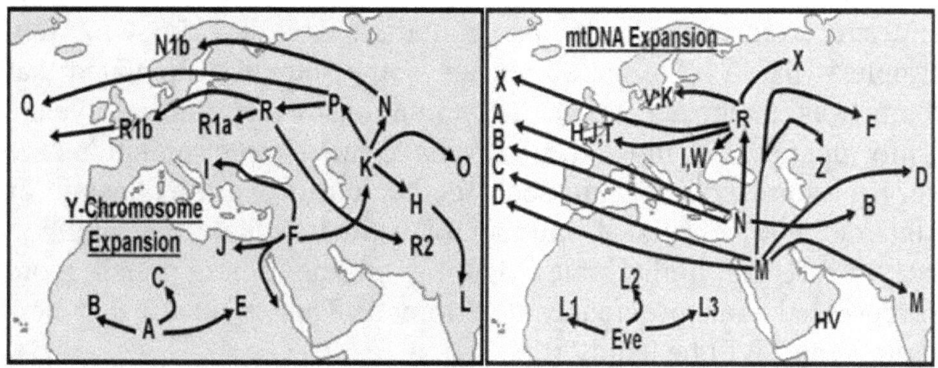

The second map is the same as the first except it is track the DNA of our special Mitochondria friends. One can define major mutation or combination points by a letter and number identification. The first letter is the most significant ancestral point. It denotes MAJOR mutation points so we can time them pretty well. The number following indicates a single event mutation/modification and a second letter identifies an additional subset or combination of groups coming together. Please note the trail of "major mutation points" is characterized by general characteristics of the groups living in those locations and how they SEEM to flow. For the example above, the string "F to K to P to R" shows a particular timeline. This is believed to be the chain of ancestry for Europeans. Of particular importance is the Haplotyping known as R1b which is the base grouping of Northern Europeans and, <u>strangely enough, Eastern North Americans</u>. Sometimes names are given to these groups to make identification easier. The dates are approximations for reference and I have added the names. The following tables show the same thing but in a table format. The first is the Nuclear or Y-DNA tracking.

Homo–Erectus
A="Y-DNA Adam" [100 thousand years ago]
B= Sapien [50 thousand years ago]
Cro–Magnon
F= Adamic [40 thousand years ago]
C= Negriod [20 thousand years ago]
D= West African [12 thousand years ago]
E= Nubian [12 thousand years ago]
G= Armenian [12 thousand years ago]
K= Japheth/Asian [12 thousand years ago]
H= Afghan [12 thousand years ago]
I= PreGreek [10 thousand years ago]
J= Hamite [10 thousand years ago]
T= Near East [6 thousand years ago]
R= Scythian [6 thousand years ago]
M= Russian [6 thousand years ago]
O= Oriental [6 thousand years ago]
P= East Europe [6 thousand years ago]
Q= American [5 thousand years ago]
S= SE Asia [5 thousand years ago]
L= Dravidian [5 thousand years ago]
N= Scandinavian [5 thousand years ago]

The second table shows the Mitochondria track again.

Homo-Erectus Humans
Eve=Erectus [100 thousand years ago]
L1= Sapien [50 thousand years ago]
Cro-Magnon Humans
N= Adamic [40 thousand years ago]
M=Arabic [30 thousand years ago]
L2= Negroid [20 thousand years ago]
L3= Nubian [12 thousand years ago]
R = Proto European [12 thousand years ago]
X= Proto-N. Amerindian [12 thousand years ago]
A=Adamic-Amerindian [12 thousand years ago]
D=Oriental-Amerindian [12 thousand years ago]
F=Mongol [6 thousand years ago]
Z=Oriental [6 thousand years ago]
B=India-Amerindian [6 thousand years ago]
C=Russo-Amerindian [6 thousand years ago]
V, K= Scandinavian [6 thousand years ago]
I, W = Greek [6 thousand years ago]
H, J, T=European [6 thousand years ago]

The Tree-Besides just getting a feeling about what mutation came first, we have details about what mutation initiated successive ones. The chart below is a much reduced version of this DNA tree. Notice that a large number of Y-Chromosome mutations occurred 12 thousand years ago and another block happened 6 thousand years ago.

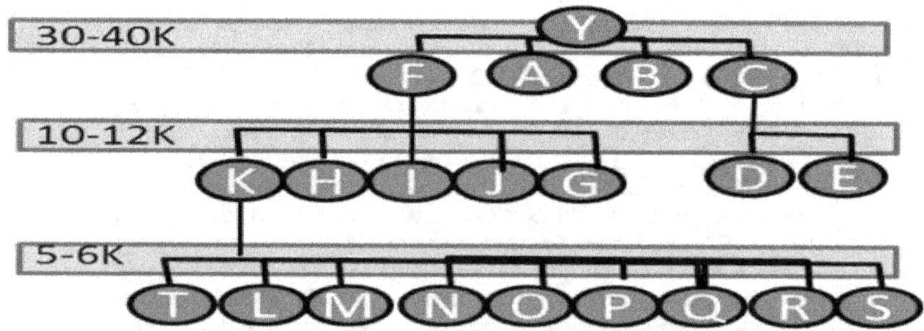

Don't worry about all this stuff too much right now. It will make sense as we go along. It should be noted that some suggest that only the "A" Genome comes from Africa as the "B" type possibly started in a different country and descendants went "INTO"

Africa. I don't know about all that and it happened before Cro-Magnon Man came along 40 thousand years ago. The mutation points before 10 thousand years ago would be those that were established before the worldwide flood. The 6 thousand year boundary would be from the beginning of the <u>Bharata War</u> or Babel War and the 5 thousand year mutation boundary was around the end of the horrible war that changed mankind forever.

That being said; please look at a couple of things in the graphic. There were 4 main types of humans or DNA mutations during the early times. While 3 were in Africa, the 4th was located in or near the area known as Iraq. All of a sudden those living in Iraq mutated all over the place as if some massive nuclear event had overtaken them. As we go forward in time, we find that those staying near Persia mutated again while those who traveled to Greece and Africa remained less mutated. In Africa, the "C" mutation people wandered northward towards Egypt and were seemingly blasted to produce more mutations than other places. By this "Tree" it seems that civilization did not "come out of Africa". Instead, the people in Africa simply stayed there.

Also notice that from Persia came the "J" mutation group to settle in Egypt. They were known as the Khemetians. Additionally, notice what happened as "P" quickly changed into "Q" mutation and was now found in the Americas as if by magic; unless one could have traveled by air or ship across the Atlantic to settle in what is now North America.

I know that was hitting you with a lot of stuff so let's look at a different testing method as scientists check the Mitochondria.

Mitochondrial Haplogrouping

While you would think tracking mitochondrial DNA would show almost the same expansion and mutation. Generally, it is similar, but there are some differences. One reason might be that the Mitochondria are more protected so there are more "straight links". Along the left of the map shows the major MtDNA of the Americas. We will have to investigate how the N and M Haplotype DNA, all of a sudden, shows up in America without secondary changes.

- **The Homo Erectus "L1", "L2", and "L3"** lines stayed mostly Africa.

- **The initial "N" Cro-Magnon** line was all destroyed except for those who survived with Noah.

- The "M" Hamite and "R" Japhethite lines move as was described in Biblical texts

- **The "R" mutation** become the Scythian just as described in ancient Irish Histories and the "B" lineage is shown to push the "M" Dravidians to the bottom of India during the Bharata War as described in other histories and the "I", "W", "V", and "K" groups begin their westward takeover of Europe finally making the "H", "J", "T" lineages the main DNA mutations of the UK.

As with the Y-Chromosome Mutations notice that a large number of mitochondrial mutations occurred 12 thousand years ago and another block happened 6 thousand years ago.

Travel to the Americas

From the map, I want you to see somehow, <u>the X,A,B,C, and D mutation groups all got to the Americas without going through Asia</u>. Many try to force the migration, but it doesn't make sense. They simply appeared. First the A and D groups followed by the B and C groups thousands of years later. It is a mystery, if no one had flying transport machines. We will look at that later. Like the Y-Chromosome map, there are many variants of these flow maps as well. Again, it should be noted that these are my names rather than ones used by others. Please notice that there are not nearly, as many "Mutations" associated with Mitochondria DNA and the Y-Chromosome so we will concentrate of Y-DNA Haplotyping mostly.

Let me show you the following chart again. As I stated, the chart shows the DNA differences between modern humans, Neanderthal, Denisovan and Chimpanzee, but maybe it means a little more now. Additionally, it provides a possible DNA difference of the Anakim people who were instrumental in developing a wide assortment of different animals and human types by gene splicing and even using their own DNA to splice into other DNA. DNA obtained from South American Paracas remains was used to generalize the Anakim details.

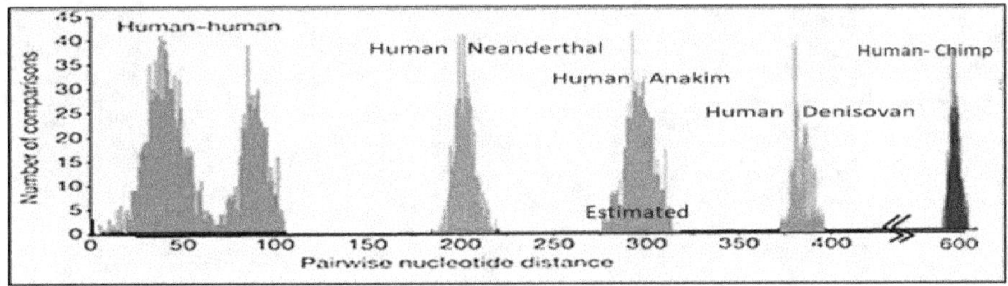

We can believe soon after the War, there were more significant differences possibly like the following:

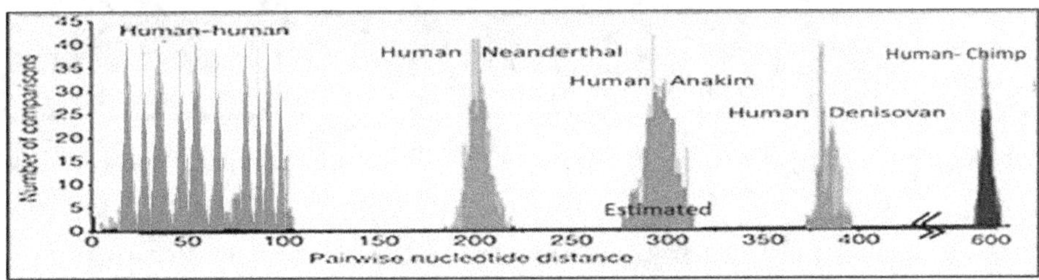

That is some basics so look at how DNA changed people over time. This first group is more ape than man, but we have to start somewhere.

Ape to Man Evolution

We find examples from Africa, Europe, Asia, India Germany, and Turkey. It is almost as if humans didn't come out of Africa at all. Some have suggested human DNA just evolved from apes who lived in Africa, but the facts don't suggest that at all. First let's look at some of the first Hominid creations. For 30 thousand years, the Anakim people created animals as required, but also many which were not supposed to be made. As can be seen here, there were genetic labs around Africa and Eurasia building animals trying to get a suitable servant. We find these ape-men were found in Spain, Germany, other parts of Europe, Asia, Pakistan, Turkey, and Africa, but we can believe there were many other locations.

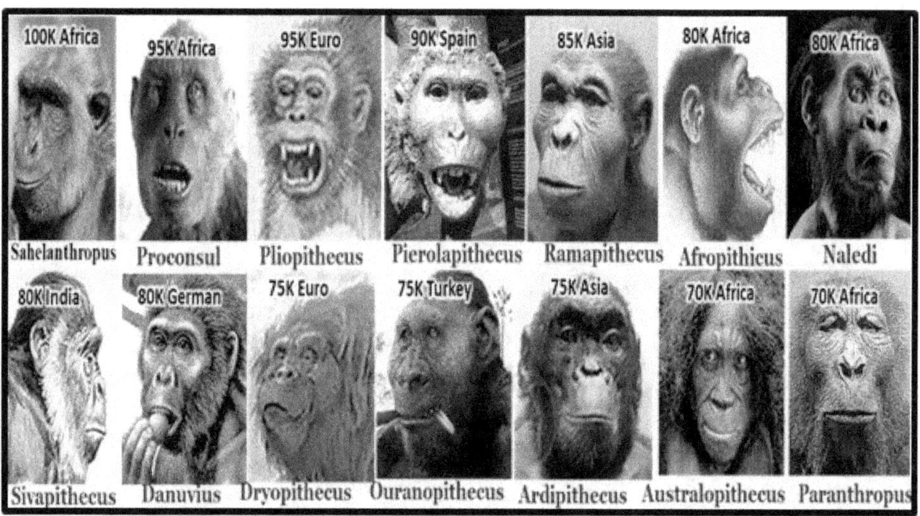

Sahelanthropus-This guy lived in Central Africa about 100 thousand years ago. Walking upright may have helped this species survive Ape-like features included a 300cm^3 brain, sloping face, large brow-ridges, and elongated skull, but very small canine teeth.

Proconsul- He was the first prehistoric mammal ever to be unearthed in Africa and he was tail-less. From the early Miocene about 100 thousand years ago; he was about 4 feet tall, standing, and weighed about 65 pounds.

Pliopithecus- This guy lived in Europe about 95 thousand years ago and was about 5 feet tall. He was similar to orangutan, but with a short tail.

Pierolapithecus- This ape-man lived Spain about 95 thousand years ago and he combined ape and monkey characteristics with a sloped face and short fingers and toes.

Ramapithicus lived in and around South-Eastern Europe and Pakistan during late Miocene about 85 thousand years ago. Similar to the Sivapithecus, this guy was orangutan-like.

Afropithecus-This ancestral hominid seems to have walked like a monkey on four feet rather two and he lived in Africa during the mid-Miocene, 80 thousand years ago. He was pretty large at five feet tall and 100 pounds.

Homo-Naledi-This guy was found in a cave in 2013, in South Africa. At least fifteen individuals who lived around 80 thousand years ago were Homo-Habilis-like hominids with long toes and fingers for climbing, but the more modern looking teeth. The brain size is of interest. At $450cm^3$, it shows this was one of the smartest of this particular group. They were 5 feet tall and weighed around 100 pounds, but they weren't as 'human' as, the much smaller, Australopithecus that would come around in another 10 thousand years.

Sivapithecus-The late Miocene primate Sivapithecus lived in India and possessed chimpanzee-like feet equipped with flexible ankles, but otherwise it resembled an orangutan, to which it may have been directly ancestral. He lived about 80 thousand years ago.

Danuvius Guggenmosi- This guy was small, about 3.5 feet high and weighed only 80 pounds, but he had straight legs for standing upright and long arms for doing ape things. He lived in <u>southern Germany</u> lived about 80<u> thousand years ago.</u>

Dryopithecus–This <u>Eurasian</u> ape-man lived throughout the Miocene about 75 thousand years ago. He was small; about 2 feet tall

Ouranopithecus- This guy lived around <u>Greece and Turkey</u> during the Miocene about 75-thousand years ago and he was pretty big at about 5-feet tall and he weighed something like 200 pounds. He was similar to the Dryopithecus.

Ardipithicus- This guy, sometimes called Ardi, lived during the Late Miocene, 75 thousand years ago, in Ethiopia. It is believed their behavior was Chimpanzee-like. He could walk on 2 legs, but could not run any distance. His brain was about $325cm^3$; similar to the Chimpanzee.

Australopithecus-Everyone knows about Lucy. She was a 4 foot tall ape-man who lived in Africa during the early Pliocene 70 thousand years ago and some DNA recovered showed she had SRGAP2 which suggests more brain function than a normal ape. Wish I had some of that as well.

Paranthropus-The most noteworthy feature of Paranthropus was his large, heavily muscled head. He is also known as Nutcracker-Man. He lived about 70<u> thousand years ago</u> during the Pliocene Epoch. He had a 500 cm^3 brain like the Australopithecus and he walked like Homo-Habilis. He weighed about 100 pounds and was about 5-feet tall.

Soon the Geneticists got better at making all types of animals, but for this book we are looking at human DNA and what it took to make it. The following graphic shows the work done between 70 and 50 thousand years ago. During this age, ape-men got huge and more ape-like to more man-like depending on the experimenter. None of the new people were much good as servants so more experimentation was required and some new blood.

Habilis to Erectus Man

This group is sometimes called the Habilis ape-men as Homo-Habilis simply means handy-man. That is what the experimenters wanted, but each of these had issue. Certainly the Gigantopithicus didn't help unless you wanted your neck broken. Almost completely Ape, this guy was 12 feet high. His cousin the Meganthropus was only 10 feet high, and he was no better. Some of this group began using simple tools, but that does not mean they helped the Anakim people build a house. You can see from the graphic these pre-humans were being manufactured around the world in China, Africa, Russia, Taiwan, and SE Asia without much success.

rectus Compared to Habilis

What we know is that there was a huge difference between the ape-man called Homo-Habilis and the human Homo-Erectus. The undirected Evolutionist quasi-Scientists can't understand what happened. It was not a massive electrical storm that enhced humaoids in opposition to Entropy. Here are some of the changes noted.

	Habilis-ape-man	Erectus-man
Location	East Africa only	Eurasia and Africa
Build	robust - 4 feet	Taller, slender -5 feet
Face	Protruding face, prominent cheekbones	flatter face, large brow-ridges
Limbs	long arms, short legs	Slender arms and legs
Walk	Feet outward	Feet more straight
feet	Hand-like	More foot-like
Posture	Stooped	upright posture
Teeth	Large and elongated	Shovel-shaped, smaller
Brain	600 cm3	1000 cm3
Tools	Scavenging tools	Hunt & defense tools
Fire	No use	fire and fire-making
Speech	no	yes

Homo-Erectus was the first true human of this line, but we find many variant of the Homo-Erectus before getting to the Cro-Magnon. His features were man-like including his teeth, pelvis, and legs. He was much larger and his brain had swelled to <u>almost twice</u> that of his very recent predecessor. Those things didn't happen by chance and they all point to a very strong, very articulate, very manlike worker. The brain expansion was almost like the brain had gone into an evolution jet going thousands of years in the future while the same general shape and appearance looked almost like a reasonable progression. Don't confuse evolution with NEW CONSTRUCTION. I know you are going to say what is scientific about that? Well the undirected evolution idea that many types of sugars just happened to be on a stump at the same time and lighning struck forcing the sugars into clumps of DNA while on another dozen stumps the exact same thing happens in exactly the same way so that a viable procreative entity could be established sounds much more fanciful. Speaking of fancy, let\s go right into Homo-Ergaster.

Habilis- was not a true man in the strict sense. Homo-Habilis lived during the Pleistocene approximately 80 thousand years ago by modern techniques. While he was called HOMO- most place him Australopithecine. I'm just putting him here for completeness. He was an advanced ape with a 600cm^3 brain. He had disproportionately long arms compared to modern humans, but he had a less protruding face than australopithecines. He made and used primitive stone tools but he had no opposing thumb and didn't have the powerful neck muscles to allow him to walk on 2 legs long.

Gigantopithicus- This guy was more ape than man for sure and he was huge at over 10 feet tall weighing something like 1200 pounds. It was almost like one of the geneticists decided to have a little fun reengineering the apes. He would have live during the time of Habilis in the areas of China, India, and Vietnam.

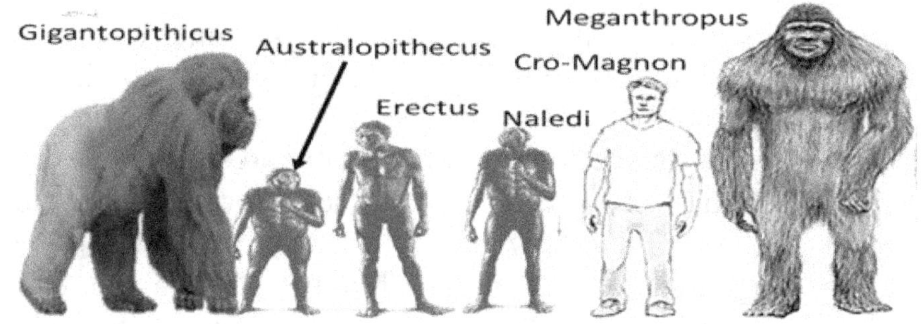

Meganthopus -After the Gigantopithicus the geneticists decided to make something slightly smaller and the Meganthropus was created. This guy lived in the same general area as the previous model but he was more human looking and only 8-feet-tall. As the skull shows, his teeth were <u>much more human</u> like, but the ridge along the brain was similar to that of today's apes. The jaw was roughly the same size as a gorilla's, but much thicker and he had a cranial capacity of up to <u>900cc</u> and he roamed the earth about the same time as Gigantopithicus. The geneticists decided to downsize and made the following about this same time.

Rudalfensis -Like Homo-Habilis this was not a human. Skeletal remains from northern Kenya. Many believe this guy was similar to the Homo-Habilis.

Georgicus -Out of Africa didn't fit so it was ignored as Homo Georgicus was found in Dmanisi, Georgia. This guy seems to be in between Homo-Habilis and Homo-Erectus. Possibly, he was human-ish and the remains appeared to be about 75 thousand years old. Some implements were used and his tooth-wear was of an omnivore. His brain was 600 cm³ and he was much smaller than Homo-Erectus he was about 4-feet-tall, with relatively advanced lower limbs.

Pangu-Man- This guy was found in Taiwan. At only 4-feet-tall he was small and similar to Georgicus with apelike arms and small brain.

Sangria Man-This Southeast Asia Variant are more Homo-Erectus than Ergaster in looks. This first guy is also called Homo-Erectus Sangria. Found in Java, Indonesia in 1969. He was a cousin to the next guy also found in Trinil, Java, in 1891.

Java-Man- Both Java and Sangria lived about 70 thousand years ago. This Erectus variant was determined by the top half of an early flat forehead skull with distinct brow-ridges and a sagittal keel [skull protecting ridge].

Ergaster -Homo Ergaster is believed to have been the first true human by some. As such, he would have been the man described in the Genesis 1:28 with the previous humanoids the ape-men identified in Genesis 1:26. As all appeared during what the Bible called the '6th age after the end of the Cretaceous Extinction described in Genesis 1:2. Homo Ergaster, was possibly the beginning of what we call the Negroid Race. Homo Ergaster simply means "working man". Without the DNA, we are only guessing at all these different types of humanoids. But we knew this one made a number of different scrappers and tools and he harnessed fire either by containment of natural fire, or as the lighting of artificial fire. He still could not make complex sounds like speech but his arms were shorter than predecessor types and he had the opposing thumb we love having and it appears have lived on the ground more than in trees. This guy has only been found in Africa and he was about 5-feet tall and 100-pounds.

Erectus -Homo-Erectus [Erect man] was found Ergaster but he lived about the same time as Ergaster. As his body shape and size was similar to Ergaster, he might have been the first true man, who knows. While their skulls are shaped differently and he was found in Southeast Asia; we can say they probably were the carriers of what has become known as the "Adam-haplotype-A". The MtDNA haplotype mutation would have been "I". These last 2 humanoids may have been able to run long distances.

Rhodesiensis -Homo Rhodesiensis also known as Homo-sapiens-arcaicus has been found in numerous sites in Africa from around 70 thousand years ago. He is considered a cousin to the Heidelberg man coming up with a brain capacity of about 700 cm^3. Strangely, he had the largest brow-ridges of any of the newer humans. He had a broad face and large nose. Speaking of strangeness, his teeth had cavities. Another strangeness is he didn't have the opposing thumb nor did he had the larger neck muscles for running. Possibly this guy was of the Homo-Habilis level.

Peking-Man- Homo-Erectus Pekinensis. Peking-Man was a Chinese Erectus with a twist. He lived about 70 thousand years ago and he used fire for cooking and heating, but he was small at only about 4 feet tall. Additionally, this guy drilled holes and no one knows why, Peking-Man was a woodworking, fire-using, spear-hafting human-ish race. As the DNA was modified the as characteristics were being removed and the Anakim characteristics were coming through.

Modification of Ape-man DNA

Guess what! Ancient people remembered something about all the modification of apes into more acceptable ape-men. Here are a few of the texts.

American Homo- Erectus Story-In the Americas, the same story was told. *"Men were once hairy and monkey like." Something happened to them.* [I think you know what it was. It had to do with Anakim and sex.]

Inca Legend-*The age of primitive man [hairy?] was before the Age of Heroes* [Anakim and demigods] [Primitive man turned into heroes probably by a process called breeding.]

Sumerian Creation Text- *After having sex with one of the Anunnaki people, the 'man' was no longer to be associated with beasts. His hair was no longer all over his body and his legs were diminished. He became a hybrid.*

India Creation Myth-*The first humans were covered with thick hair. When they mated, they produced people as they are now.*

Sumerian Gilgamesh Story-*Aruru [an Anakim] pinched off some clay and created a [primitive man] Enkidu. His whole body was covered in hair. With cattle, he quenched his thirst, Shamhat [an Anakim] she did for the primitive man [Homo-Erectus], as women [Anakim women] do. She pulled not away. Afterward- the gazelles saw Enkidu [offspring of] and scattered, for Enkidu's body was hairless. His legs were diminished; he could not run as before. He had become wiser.*

Mandaean[12] History- *When man was finished, he looked like a man, but moved on all fours, had the face of an ape, and made noises like a sheep. Only later did he put in a soul and teach him and make him erect.*

---

*12 **Mandaen**- They occupied Iran during an ancient period. This version of the Genesis story is probably only about 2 thousand years old. In this Mandaean version of the creation of the 6th 'day' man, we find that it was not a perfect "creation". Just as implied in the Biblical version, the "human" had to go through some genetic changes before it could be perfected. Initially, he was an ape-man.*

Southeastern African Verbal History- *Hairy men became human after coming in contact with Annu [Anakim].*

Southeast Asian Tradition *-An extremely hairy human "female" named Bota Ili was cooking food. A non-hairy fisherman named Wata Rian [an Anakim] saw her and got her drunk. While she was asleep, he shaved her entire body. Only then did he find out she was a woman. She learned to wear clothes, they married, and they began a new race.*

Ngombe Tribe[13]- *A sky person [Anakim] saw a hairy man [Homo-Erectus]. She married him and removed his hair. Then a Garden was made for man to live in.*

India Creation Myth-*The first humans were covered with thick hair. When they mated, they produced people as they are now.*

Gilgamesh Story[14]-*This seed, created races of men from around the world. From the seed of a fallen god [Anakim/Anunnaki] she made mankind.*

From here, we begin the effort to make human brains bigger. The map following shows the main areas the preceding group and the big brain group coming up lived. There was activity in the Americas, but let's concentrate here as similar DNA modifications were done everywhere.

---

*13 **Ngombe**- [means water people]-These people generally lived around the Republic of Congo and were fishermen, hunters, and cannibals. Today only 150- thousand speak the language, but it may have been much larger when these stories were first told and retold.*

*14 **Gilgamesh Story**- This Babylonian record is similar to the ancient Sumerian version. This much newer 3200-year-old version cross-compares to the Biblical account. This section reinforces the Biblical Anakim people as the losers of the Heaven War and the hybridization of the Cro-Magnon people, and the survival of some hybrids.*

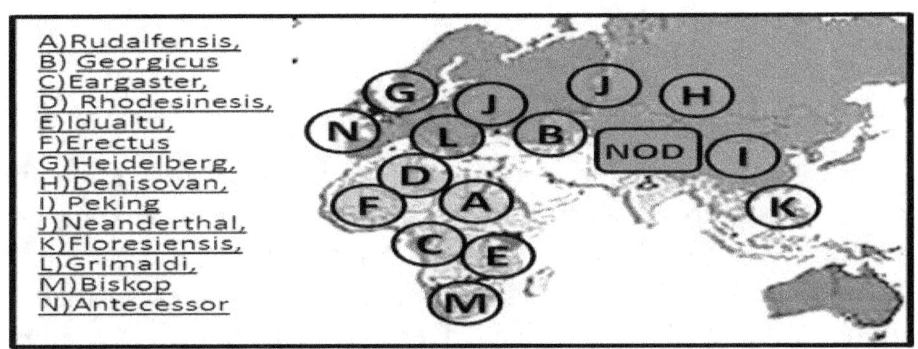

The indication of 'Nod' shows where the Biblical detail described the first place, Cro-Magnon could have been modified by gene splicing or sexual intercourse with the Anakim people to make mixed hybrids. New "races" were showing up every year as awards might have been given out for the most creative developments. We can imagine many didn't make it. They might have had only one eye, or, slither like a snake. Some would not be human at all. Many probably were killed while most probably just died as the DNA could not sustain life. They wanted bigger brains.

Bigger Brains

It wasn't long before the Anakim biology labs were adding in Anakim DNA into the early humans that had come into existence during the 6th Age. The brains would grow from 700cm^3 to about 1600cm3 between 70 and 40 thousand years ago as experiments continued and DNA was spliced. Here are some of the versions found to date.

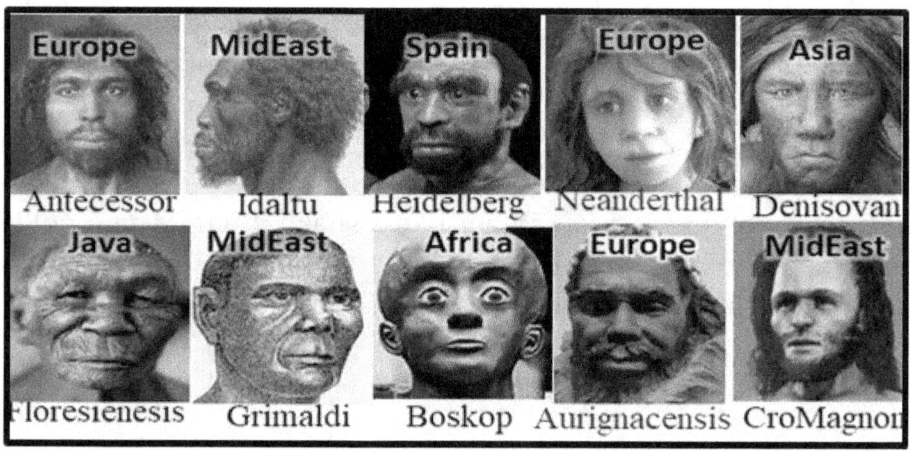

Antecessor-Man -Called Homo Antecessor, these people lived well before Neanderthal sometime around 70 thousand years ago, in Europe. Here is the bad part. They were cannibalistic; Antecessors ate Antecessors. If one invited you over for dining, it is suggested that you refuse. These people were full sized at about 5½-feet tall, and males weighed about 200 pounds. Like I said, their brain had jumped to roughly 1,000 cm^3, about 25% smaller than our modern brain, but still smart enough to fillet a side of Antecessor. Also it is believed these were the first to use language. They still had the protruding brow ridge, low forehead, and a lack of a strong chin but there seemed to be many of them with finds in Spain, UK, and France. The Anakim people has successfully converted Homo-Erectus into this first modification and they kept eating each other, so they continued.

Idaltu-Man-This may have been the next experiment, living around 65 thousand years ago, in Africa. "Idaltu" just means "first born". This time the brain had increased to a little larger than our modern brain, at 1,400 cm^3. Considered the oldest anatomically modern humans, he was a tool maker and more than 1000 stone artifacts, including hand axes, flake tools, cores, flakes and rare blades. Here is an interesting fact. They seemed to like eating Hippos.

Heidelberg -These people were called Homo Heidelbergensis and they lived in Europe well before the end of the Pleistocene and before the Neanderthal. They have been found with Antecessor remains so there is no telling what conflicts might have arisen. We can imagine they did not like being eaten. This guy looked much like a modern man.   It is believed these were the first civilized humans who died off about the time of the Cro-Magnon we will look at later. Mostly found in Northern Spain, but some have been found in England and Germany. Their brain was in-between the Idaltu and Antecessor or similar to modern humans. Like the others, they were about the size of modern humans as well. Here is an important fact. Instead of hippo they ate prehistoric bear and vole. They also buried their dead---possibly so Antecessor would find an easy meal. Hundreds of hand-axes made by these people have been found in England and Spain and 8 wooden throwing spears were found in Germany along with 16 thousand animal bones. From this German site it is believed these guys were very good hunters.

.

Denisovan-Let me, let you in on a dirty secret. If you want to have a viable community, you must have many opportunities for procreation attempts of DNA. Some don't work, sometimes location is a huge deterrent, sometimes the offspring are not viable, etc. etc. That being said, tracing back family lines to a single group from a single parent is not likely. I'm not saying there will not be similarities in how each of the potential offspring producers would mutate, but to think that all would mutate the same is--- OK! I'll say it again; UNLIKELY. Somehow, the Denisovan Race with a brain even larger than the later Cro-Magnon and over 30% larger than modern humans, didn't get this information and messed up a number of Haplotype studies that didn't rely of ancient Laboratory, Genetic modification. Denisovan people seem to be a cross between Heidelberg and Neanderthal people; in fact, they are more closely related to the Heidelberg people according to their Mt.DNA. A fairly good Denisovan genome revealed variants of genes in humans that are associated with dark skin, brown hair, and brown eyes showing Denisovan looked substantially different than Neanderthal. Later we find the strange Paracas people were genetic relatives to these impossible people.

Floresiensis –Not long after Denisovan was established, a group of tiny people at only 3 ½ feet tall and weighing about 60 pounds, were manufactured. It seems they were only found on the Island of Flores, Indonesia where we find something very curious. We find they hunted tiny elephants, huge rodents, and Komodo Dragons. If you ask me this experiment was sort of a zoo for back then. Visiting the Island would give the Anakim people of the Pleistocene Age some comic relief or it may have just been an experiment of some kind. From limited DNA details, the strangeness seems to show the species is not associated with others. The images show the size of their skull, one of these guys carrying home a rat, battling, and the comparison of their tiny elephants. I think the biologists were just having fun with DNA.

Grimaldi man- This next human was Negroid in features and had a brain that had grown to about the size that would be established by Cro-Magnon man. At 1600cm^3 it was about 20% larger than our modern brains.   This extremely intelligent human species found in Italy, was little more than 5-foot tall, more slender than. The faces had wide nasal openings and <u>lacked the rectangular orbital and broad face so characteristic of Cro-Magnons</u>. The nasal bones gave a high nasal bridge. Fine artwork has been found in caves occupied by these people. One such artwork is shown below.

Boskop man- We find a similar large brained man back in Africa. Known as the Boskop Man because he was found in Boskop, South Africa, these people must have been "worked" on by the Anakim scientists. With a large brain and classified a Homo-Capensis, his brain was close to 30% over a modern human so this guy must have been something. Similar to Grimaldi, their narrow faces had wide nasal openings and lacked the rectangular orbital of Cro-Magnons.

Neanderthal- That brings us to one many know, called Neanderthal. Neanderthal as a race disappeared about 15 thousand years ago and came into existence about 60 thousand years ago and we have been able to sequence the DNA. With about 15 partial MtDNA sequences to test, scientists have found something interesting as they tested samples from central Asia and all over Europe. There are very few similarities in Neanderthal and modern races and there were, possibly, as many as 4 different sub-races of Neanderthal and he had <u>alien sections</u> that did not seem to come from apes or modern humans. It was as if a different human gene spliced his own DNA into this new man. He had a massively thick skull over a very smart, brain that was 10% larger than our modern brain. According to DNA, his complexion was light, he had reddish hair, sang with a tenor voice and had some type of language he used. Yes; he had a big nose, but he probably liked it that way. Researchers tell us he possibly had a 70 word vocabulary and carried the "ginger Gene" that makes people have freckles, and the reddish complexion and hair. Evidence of wearing clothes and shoes tools, hunting skills, burials, handmade jewelry, musical instruments, toys, and other things allow us to know this was a well-developed and social group of people who lived in circular homes that used mammoth bones for wall supports. Here is the interesting thing. <u>Researchers have found that Neanderthal DNA had key elements of pro-creation missing in the X-Chromosome DNA. This suggests that they must have had difficulty in having children, which would have frustrated the Anakim scientists.</u> Sometimes, I think Neanderthal DNA has been tested more than modern man. I know that's not the case, but here are some examples.

Testing Neanderthal DNA

As scientists started looking at DNA from a number of sources we started to get a more definitive accounting of what Neanderthalis was like just by reading DNA codes and comparing them with our modern DNA. As I mentioned, we now believe Neanderthal were light skinned and red-haired singers with a tenor voice.

Red-Headed Neanderthals-Ancient DNA has been used to show aspects of Homo-Neanderthalis appearance. A fragment of the gene for the melanocortin 1 receptor (MRC1) was sequenced using DNA from two Homo-Neanderthalis specimens from Spain and Italy in 2007. Neanderthals had a mutation in this receptor gene that <u>has not been found in modern humans</u>. The mutation changes an amino acid, making the resulting protein less efficient. Modern humans have other MCR1 variants that are also less active resulting in red hair and pale skin. The less active Homo-Neanderthalis mutation probably also resulted in <u>red hair and pale skin</u>, as in modern humans. Because this (MRC1) stuff was in Neanderthal and not in us, this gene came from someone else. Many scientists struggle with this because they will not accept the existence of the Anakim no matter how much evidence shows up.

Speaking and Singing- Another odd sequence called [FOXP2] was found. The FOXP2 gene is involved in speech and language and this FOXP2 gene is mutated in Chimpanzee, but not in Neanderthal. Therefore it is believed Neanderthal did not just grunt when he saw a good looking woman. He might have said something about her eyes or shape rather than clubbing her in the head as we once believed, but recent physiological discoveries indicate that their voices were high pitched and nasal, not the baritone grunts we normally associate with cavemen. No one has found any of their lyrics, but I made one that might have been.

I'm not a caveman, but I paint in caves-

The next verse is mostly humming, so let's continue by looking at blood.

Blood type- Interestingly, researchers found some Neanderthalis people had type O blood. Possibly it was OA or OB, but nothing crazy.

Microcephalin gene –This one has been in the news lately as the news media claimed Zika virus would shrink the heads of babies making them Microcephalin. Researchers have determined that this gene comes from the Haplogroup D. Initially, it was believed Neanderthal carried this problem gene, but later studies disproved it. One can still believe Microcephalin genes came from a Homo-Erectus that was not Neanderthal as Neanderthal had big brains.

Comparison to Chimpanzee- As I showed earlier, researchers compared the Neanderthalis to modern human and chimpanzee sequences. Modern human sequences varied between each other with about 8 substitutions in the DNA chain, but chimpanzee sequences by about 55 substitutions when compared to the mean of modern humans and Neanderthal had about 27 substitutions, so by this 1997 study and another in 2000 we could say Neanderthal was half way between Chimpanzee and Modern man. Chimpanzee also has a closer matched DNA structure to humans than any other Ape. When comparing Chimpanzee and Bonobo, Pan-Troglodytes, have about 1% difference in DNA also it was determined that Chimpanzee diverged from humans only about 6 thousand years ago and Bonobo may have been slightly less than that.

When Did a Differenced happen?- Researchers found modern human and Neanderthal human remains from about the same time frame and found that the modern looking human had modern DNA sequencing while the Neanderthal had sequencing similar to the other Neanderthal. This showed that early anatomically modern Homo sapiens were not very different genetically from current modern humans, but were still different from Neanderthals. It was as if Cro-Magnon was a new species that simply appeared and didn't change much since it "appeared".

Europeans and Neanderthal- Many Neanderthal remains have been found in Europe and European skulls are slightly more Neanderthal-like that other parts of the world so researchers believed that would finds a link but there has been found no closeness of Europeans than any other modern human. Also various analyses have examined the amount of Homo-Neanderthalis contribution to modern human mtDNA. The analyzers were unable to find positive evidence for interbreeding between modern and Neanderthal humans. It was as if the group that became Europeans, settled there after the Extinction of the Pleistocene Age and had nothing to do with who had been there before. This is an important point as many try to force fit colonization by location of ancient remains from before the Pleistocene event. No wonder there is so much confusion, I will discuss this more later.

A 2002 study concluded- *Homo-Neanderthalis and modern human mtDNA is consistent with large-scale replacement and some small amount of interbreeding between modern and archaic [Erectus] populations. Interbreeding between archaic and moderns may have involved different species of archaic humans, including populations in Africa, Asia and Europe.* It showed more similarities between non-African modern humans and Neanderthals than between African modern humans and Neanderthals but all differences are very minor. Approximately 2.5% of non-African modern human DNA is shared with Neanderthals while Africans share only 1.5%.

No change- A study in 2009 indicated that Neanderthal changed only about 1/3 as much as the changes noted in Modern human DNA. This doesn't make sense in that Neanderthal supposedly lived for a longer time period. If researchers would simply get off of this nuclear decay timing, they could understand that modern humans have been around longer than Neanderthal. Using nuclear decay timing it was determined that the spit that resulted in modern humans from Neanderthal occurred about 450 thousand years ago, and their heads were about to explode. One would expect with modern humans being on Earth 40 thousand years and Neanderthal living over a period in excess of 400 thousand years. Using the newer dating methods, Neanderthal only existed about 50 thousand years total so the tiny number of changes makes a lot more sense.

Neanderthal Location Mystery-One mystery seems to be how people got where they were. Neanderthal DNA contribution has been very scarcely found in African populations, but there is an exception when testing the non-African DNA portions of the Maasai, in East Africa. Not only is there enough similarity to show contact, DNA-ologists can tell when the contact occurred. It can be concluded that recent non-African gene flow was the source of the contribution as about an estimated 30% of the Maasai genome from about 100 generations ago. [2500 years ago- not 100 thousand years ago as some had speculated.] All that being said; a new human type showed up before the time of the Cro-Magnon rise to power. It was good old Neanderthal. It just appeared as if a scientist modified DNA to augment the Homo-Erectus people.

Complete Sequencing of Neanderthal DNA-Let's look at the DNA charts from before in a little more detail. They seemed to show how very close Neanderthal and modern humans were. This first graph developed in 1999 by Dr. Krings and his DNA scientists shows distributions of sequence <u>differences</u> among Humans, the Neanderthal, and Chimpanzees. This is what they stated. *It may be noted that a small fraction (0.04%) of the inter-human comparisons are larger than the smallest distance (29 substitutions) between the Neanderthal and humans.* In this initial work seemed to show Neanderthal could be considered a modern human.

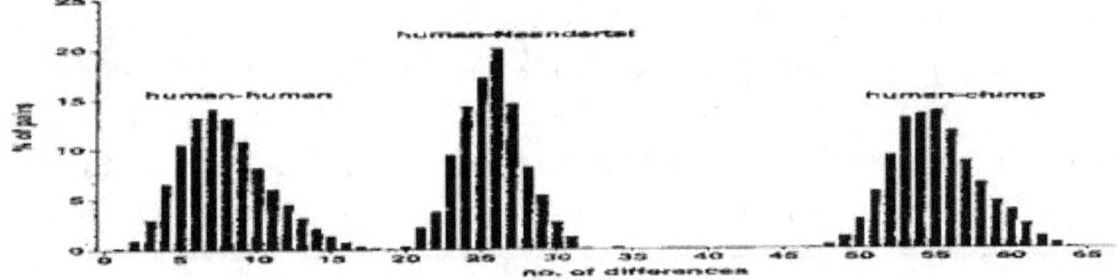

As I said it was found that this was not a correct assumption. In 2008 A team under Dr. Green were able to sequence almost all of the Neanderthal DNA. The chart below is from his study which not only looked at numbers of differences, but also where the traits were positioned in the DNA sequences. It not only shows there are 2 types of modern humans, but also that there was a wider separation than originally was believed. Neanderthal had very small differences to modern humans and much closer than chimpanzee.

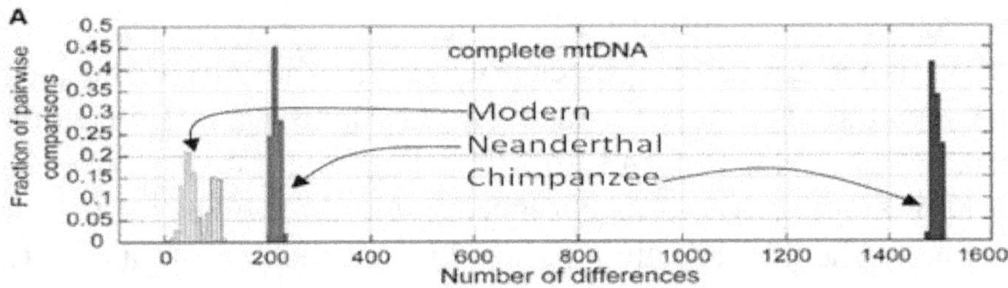

In May 2010, Dr. Green provided us the first Neanderthal nuclear DNA scan of about 2/3rds of the entire thing from 3 specimens which show that Neanderthals interbred with humans, and that all non-African modern humans contain 2.5% of Neanderthal genes. They also suggested that because of the high levels of Neanderthal traits in Asians that Neanderthal probably mixed with Cro-

Magnon in the Middle East as almost NO Neanderthal characteristics are found in African peoples.

Where are the People with Neanderthal Genes?-While we found most Neanderthal remains in Europe, testing modern humans describes something strange. The following map shows where people live that have more Neanderthal DNA traits that the rest of us. This includes large tracks of North, Central and South America, the Far East and Oceana. [I circled the areas of concentration so they would be easier to see.] There are almost no people with Neanderthal traits in Africa, Australia, and Europe.

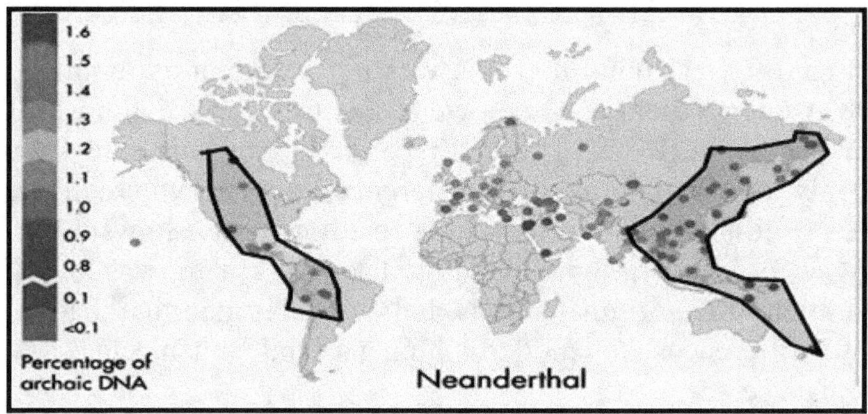

Who Were the Neanderthal?-As this will be important later, as scientist have run full sequences of Neanderthal DNA and partial DNA details of a second race called Denisovan. Both looked similar to Cro-Magnon and both had light complexions, red hair, and tenor singing voices as his larynx would have been articulated like ours. The Neanderthal had a brain capacity of about 10% more than modern man and the Cro-Magnon had a brain capacity of about 20% more than modern man so we can assume both were much smarter than modern humans. Sometimes anthropologists forget to tell you about that. This detail will be very useful as we track ancient South American overlords, as they had both Denisovan and Neanderthal trait-indicators from before the Pleistocene Extinction.

Few Neanderthal in USA-Additionally, the Adena people of North America had a Scythian DNA heritage (haplotype R mutating to X) and the preInca of South America had a Middle East DNA heritage (haplotype N mutating to A), both, evidently, showed up around 8 to 10 thousand years ago while the preMaya of Central America had the heritage of both the PreInca and another haplotype indicator that that <u>came from India</u> (haplotype M that mutated to B) which happened about 5 to 6 thousand years ago.

Denisovan DNA-Somewhat similar to Neanderthal, a Denisovan bone was found in Russia, the more interesting thing is that this seemingly close relative to Heidelberg and Neanderthal who both were European had its highest concentrations of similarity to the Australian Aborigines.

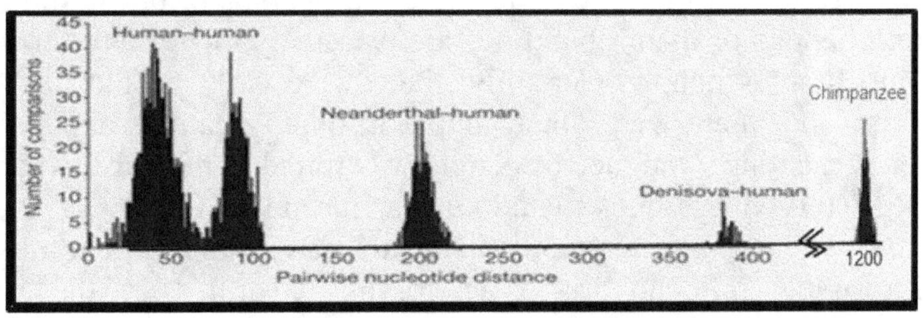

Today we find very few Denisovan specific traits, but where we find them may be telling as there are none in Africa, North America, and northern Asia. Don't worry about this map too much as the maximum similarity of Denisovan is about 0.5%.

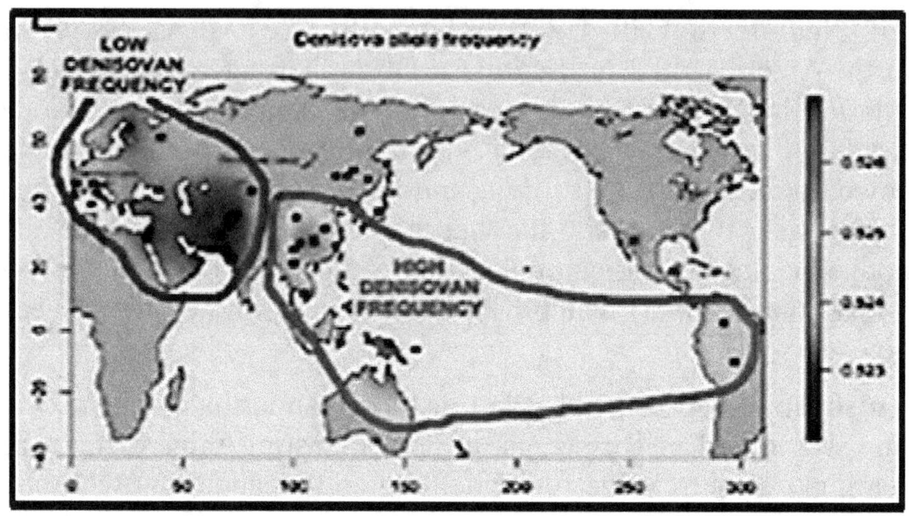

Not Out of Africa-I know you keep hearing a new mutation of people came out of Africa 200 thousand years ago, but it simply didn't happen or the probabilities are extremely low except for the Homo Erectus people. Certainly, there were early people in that country, but there were other people in other locations as I have been presenting. One race of people was found in the Far East and SOMEHOW these guys mated with a European Heidelberg from Spain. The researchers reconstructed a nearly complete genome of this fossil's mitochondria. The fossils unearthed at the site resembled Neanderthals, so researchers expected this mitochondrial DNA to be Neanderthal but to their dismay, the ancient Mother was Denisovan. <u>Now it seems that these Denisovan stayed with Neanderthal and went northwest to get to Spain</u>. Then the Denisovan group was mutated without the Neanderthal getting exactly the same mutation and most of the Denisovan left for Australia while at least the finger of one of them went north to Siberia. Please don't get me wrong. I don't believe this story. It is what they are trying to make everyone else believe. From this, let's go to something we have details about. We know quite a bit about the beginning of the "8th Age" and the "creation of Adamic people". Scientists call them Cro-Magnon or

Homo-Sapien-Sapien. Don't ask me why Sapien has to be written twice.

Artificial Mutation-For the average person about 4% of your DNA is known to be from Neanderthal; whereas, about 5% of the genome of Melanesians DNA came from Denisovan. Tibetans have a DNA Haplotype mutation that assists with adaptation to low oxygen levels at high altitude and in 2014 it was found that <u>Denisovan had the same mutation</u> showing their close relation. With the massive distances between each of those that seemed to be connected with Denisovan, it is easy to conclude that the Anakim were working to modify and make Neanderthal more rigorous. One group of Anakim must have tried their genetic art in Indonesia. Neanderthal and Denisovan were wondering around and then there were Cro-Magnon.

Cro-Magnon

I know scientists have tried to tell us there must have been widespread inter-marriage and cross-breeding, but when one researcher was asked, his answer seems more reasonable. He said not likely--

- *They were probably physically repulsive to each other*
- *They couldn't meaningfully communicate*
- *Beer wasn't invented yet.*

With that here are details about a brand new subspecies. Cro-Magnon seems to have just appeared one day. The pure Cro-Magnon were Haplotype "F" and they carried Mitochondria with Haplotype "N" if you were wondering. As for Cro-Magnons, they're pretty much like us except for their 1600 cm^3 brain. Compared to a Neanderthal, they had broader faces, a bit more muscle, and larger brain. They made what we call complex tools rather than simply knives. They spoke and <u>probably sang more often than Neanderthal and no, I'm not presenting a song here. . They lived in more well-built</u> homes, made jewelry, used burial rituals, and came up with a calendar and something very interesting. They wove cloth and skins for clothes, wove a variety of shoes and sandals. Here is just one report of a Cro-Magnon find under Death Valley Nevada. This comes from the August 5, 1947 edition of the San Diego Union Newspaper article entitled *"Atlantis in the Colorado River Desert"*. The article is shown next followed by an overview that is easier to read.

Trace of Giants Found in Desert

LOS ANGELES, Aug. 4 (AP)— A retired Ohio doctor has discovered relics of an ancient civilization, whose men were 8 or 9 feet tall, in the Colorado desert near the Arizona-Nevada-California line, an associate said today.

Howard E. Hill, of Los Angeles, speaking before the Transportation Club, disclosed that several well-preserved mummies were taken yesterday from caverns in an area roughly 180 miles square, extending through much of southern Nevada from Death Valley, Calif., across the Colorado River into Arizona.

Hill said the discoverer is Dr. F. Bruce Russell, retired Cincinnati physician, who stumbled on the first of several tunnels in 1931, soon after coming West and deciding to try mining for his health.

MUMMIES FOUND

Not until this year, however, did Dr. Russell go into the situation thoroughly. Hill told the luncheon. With Dr. Daniel S. Bovee, of Los Angeles—who with his father helped open up New Mexico's cliff dwellings—Dr. Russell has found mummified remains together with implements of the civilization, which Dr. Bovee had tentatively placed at about 80,000 years old.

"These giants are clothed in garments consisting of a medium length jacket and trouser extending slightly below the knees," said Hill. "The texture of the material is said to resemble gray dyed sheepskin, but obviously it was taken from an animal unknown today."

MARKINGS DISCOVERED

Hill said that in another cavern was found the ritual hall of the ancient people, together with devices and markings similar to those now used by the Masonic order. In a long tunnel were well-preserved remains of animals, including elephants and tigers. So far, Hill added, no women have been found.

He said the explorers believe that what they found was the burial place of the tribe's hierarchy. Hieroglyphics, he added, bear a resemblance to what is known of those from the lost continent of Atlantis. They are chiseled, he added, on carefully-polished granite.

He said Dr. Viola V. Pettit, of London, who made excavations around Petra, on the Arabian desert, soon will begin an inspection of the remains.

"Near the Nevada - California - Arizona border area, 32 caves within a 180 square mile area were discovered to hold the remains of ancient, strangely costumed humans. These Pleistocene individuals had been laid to rest wearing manufactured clothes fashioned into medium length jackets with trousers extending slightly below the knees, similar to grey sheepskin. They also found stuffed animals including Elephants and Tigers.

Can you imagine stuffing the skin of an elephant? Maybe using one of the blow-in foam things they use to package mail would have been good. Anyway; this article didn't describe the footwear, but we are finding shoes and imprints of them all over the place from Pleistocene times and even before. Some found were leather and some had hard materials on the bottom. Specimens of a few are shown next. The first specimen even has the fine stitching of multi-fiber thread. The four in the middle are a small portion of the100+- comfort, thatch-braided, varieties, found showing most people wore shoes during the Pleistocene and early Holocene. The last shoe impression shows this ancient human stepped on two trilobites, before trilobites became extinct. Maybe these were the last 2.

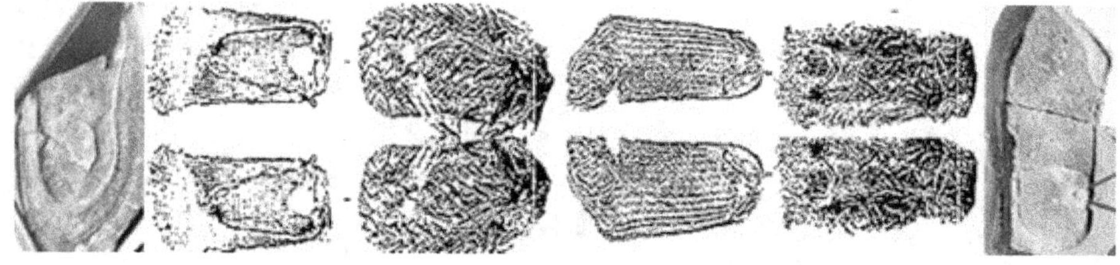

We can believe, all the humans with similar and larger brains than our current brains knew that wearing shoes was a good thing. This would include Neanderthal, Idaltu, Boskop, Denisovan, Cro-Magnon, and other large brained men and women. A wide variety of footwear and impressions of some have since been found all over Europe, Asia, America, and the Middle East. From details at the finds, the locations, and bones it has been determined that many Cro-Magnon lived a life similar to that described about the war-hungry Scythians with violence and constant physical activity outdoors. This developed them as an athletic, graceful, and gymnastic human type. They buried their dead even when the fear of being eaten was not a major concern and they cared for and protected wounded family members. Cro-Magnon were like us except for their much larger brain.

We know the story about the first Cro-Magnon in Genesis 2 as God placed him in the Garden of Eden. Please understand this outside help by the Anakim and God using what we can call "controlled Evolution" follow DNA modifications much more closely than all that impossible uncontrolled evolution using the Law of Entropy. Anyway, the location of this Eden Place was described as shown, and, sure enough, guess where Cro-Magnon appeared? Right near the Garden which would be destroyed along with everything else after another massive war.

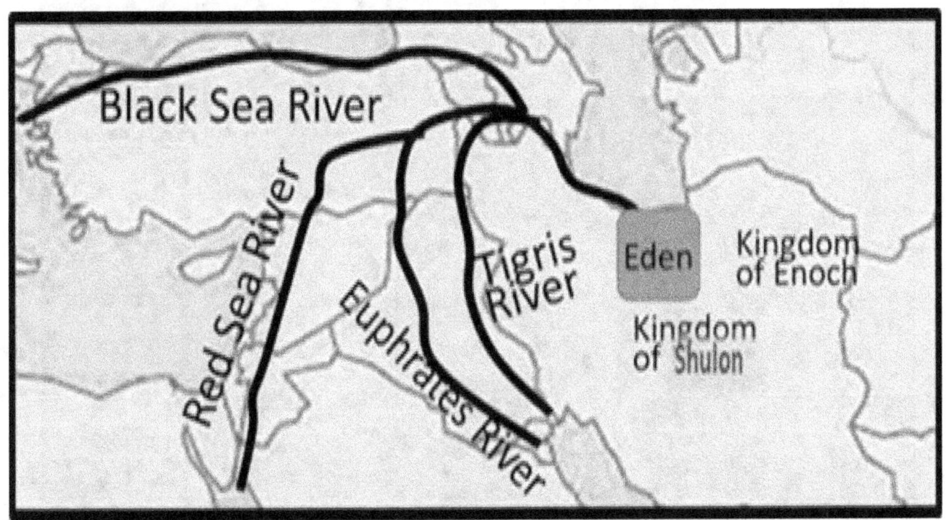

Somehow caught up in War-We are told, wars were often and bad during the Pleistocene Age. Many ancient texts describe them and some physical evidence was even found and hopefully I have given you the information you need to begin to sense the DNA modification in the Pleistocene Age, but we also need to be looking at war. As war began, some of our "cousins" were found to have been shot. I know you were told the Cro-Magnon possibly preyed on the Neanderthal and eventually eliminated the species, but there is much more to the story. Anakim, Anunnaki, lords of Amenti, or Arya, depending on the society describing their ancestry, were in control of the world and war was initiated to keep their control and hold power.

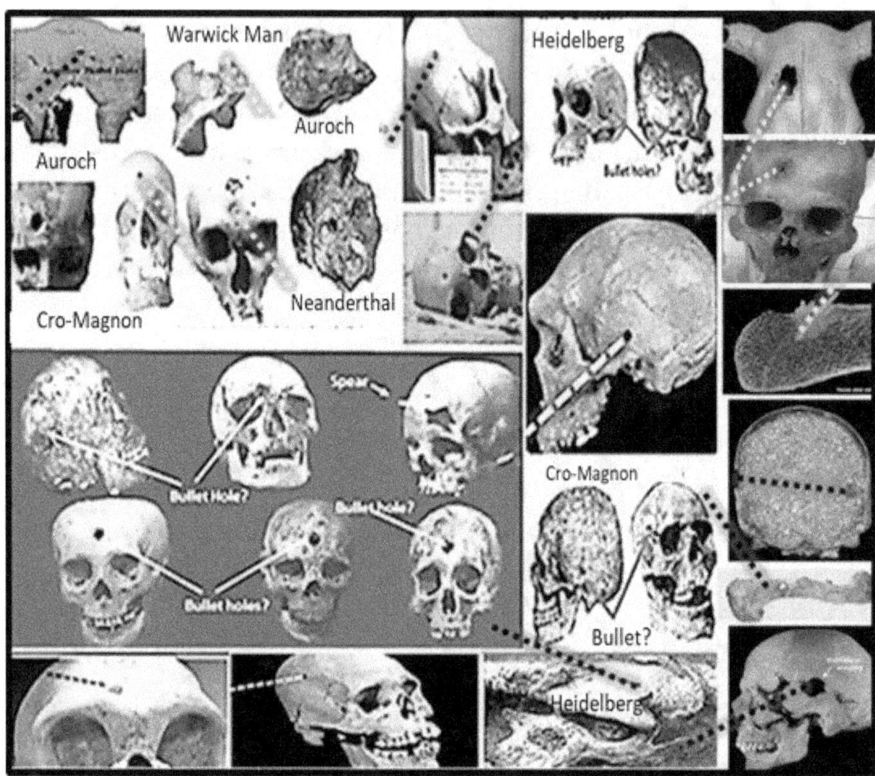

War was everywhere during these ancient times. Just reviewing the projectile kills shown in the graphic we have 3-Auroch buffalos, the Kennewick man, 3-Paracas humans, 2-Heidelberg men, 6-Neanderthal, and about 11-Cro-Magnon humans that were caught up in some type of events leading small, high-speed projectiles giving them a bad day. Two projectiles were spears, but most are smaller, more deadly circular holes. It seems like we find more each year. And modifications to Cro-Magnon seemed to have gone crazy between 10 and 12 thousand years ago.

Modification Of Cro-Magnon

As the wars began, we can imagine the scientists pushed to make new human-soldiers. The Germans tried doing this by inbreeding certainly people with good traits, but the Pleistocene scientists added the modification of sections of DNA just like they did with the lesser breeds of humans. The new Cro-Magnon were very close to the Anakim people and after a little bit of "conditioning", some of the Cro-Magnon could procreate with Anakim.

Enoch[15] 2:18-Three [Anakim] came down and copulated with women and had offspring. [The "three" probably indicates three successive attempts at inbreeding with humans or three simultaneous inbreeding attempts.]

Generations of Adam[16] 1-Eight times since our leaving the garden had the moon become new, when a son was born unto my wife Lilith. We all rejoiced at this appearing and on the eighth day when I dedicated him unto Jehovah, I called him Cain

Zohar[17]---1:19-When Cain was born, Lilith could not attach herself with him, but later she approached him and bore.—

As soon as the Anakim scientists found that the Cain-ite descendants could accept portions of the Anakim DNA, all bets were off. There were going to be new humans made. Some hybrids were made the old fashioned way.

---

15 Enoch- This sacred work was so revered, we have found 13 copies among the Dead Sea Scrolls and it is also in several modern Bibles.

16 Generations of Adam-As one of the most sacred scriptures of the early Christians and described inn Genesis 5, this book is still found in the Messianic Essene Bibles today.

17 Zohar-These are the main writings of the ancient Jewish Cabbala written about 150 AD. While there is some mysticism ascribed by the Early Ebionite Christians that used Zohar, it adds affirmation to other works.

***Melchizedek**[18]-Pray for the offspring of the Anakim, together with seed which flowed forth from the father of all who made the entire universe from nothing there were engendered the gods [ANAKIM people] and angels, and the men that came out of the seed, all of the natures, those in the heavens and those upon the Earth—now the nature of females was wanting among those that are in the heavens. They were bound with men and women, but <u>these were not the true Adam</u> nor the true Eve.*

This verse talks about a difference between angels and people called the ANAKIM and infers that a union between man and one of Anakim was accomplished. It specifically indicates this Hybrid or Gentile human was not the true Adam and Eve.

The world was changing faster and faster. War was getting bad, but societies continued. As the Anakim of this time were still larger than "normal" people we still find larger weapons a tools from this time as well as intricate manufactured items made from exotic metals and other materials. Titanium and Aluminum were processes, Tungsten was used in designs. Engines ran on some type of mercury vapor and electricity apparently was becoming common. Soon, the war getting out of control. We are told massive weapons were created and used and colonists on Venus, called Rahab in the Bible, got involved. It would be the destruction of Venus that may have actually initiated the ending of the Pleistocene, but whatever caused it, the Earth shifted on its axis by over 180 degree. This violently ended one Age to produce the Holocene.

Of all the advancements of this time, the one that was responsible for the most advancement and destruction at the same time was air travel. We read stories of colonization, especially on Venus, battles on the moon and space stations circling our planet.

---

*18 **Melchizedek**- This sacred work was part of the Dead Sea Scrolls and in the Ethiopian Bibles.*

Anakim Controlled the World

Long story short, the Anakim people lived during this time and controlled just about everything. Cro-Magnon were not the highest lifeform and the Anakim continuously reminded the normal people that they were like gods. The following shows a small sampling of the many articles that have survived over 10 thousand years.

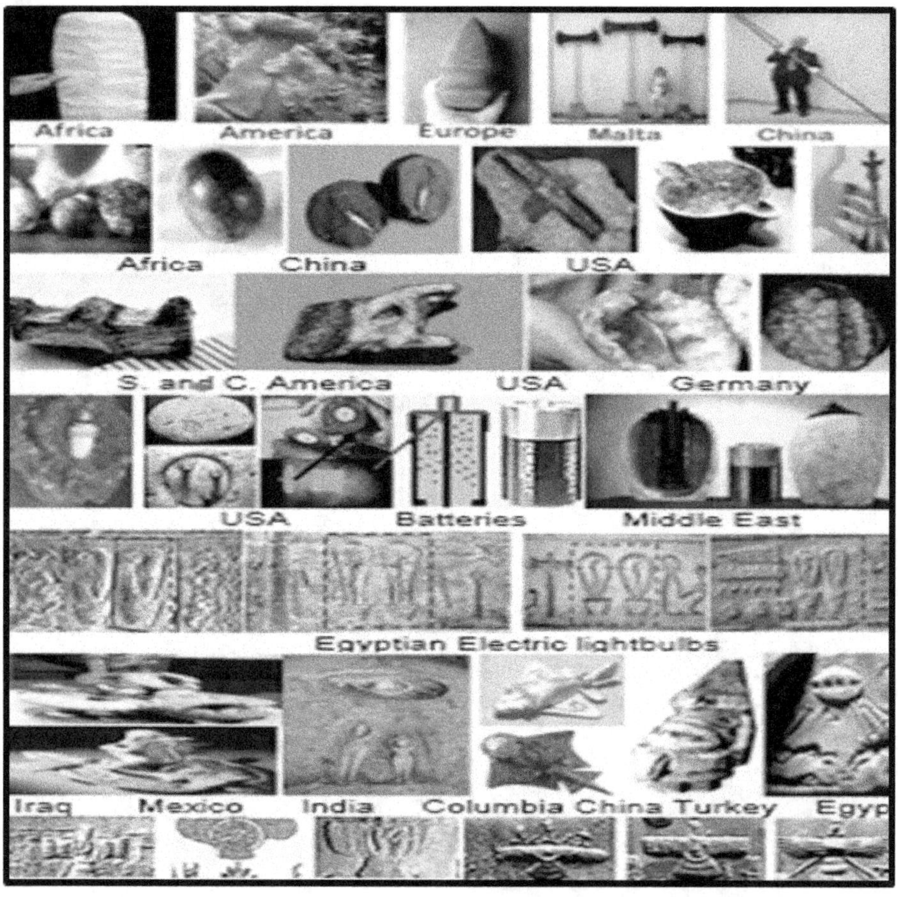

The first rows simply shows how larger the first Anakim were as their tools and weapons were huge as well, the Second row shows cannon balls, projectiles embedded in rock, a hammer and other manufactured goods, the third row show exotic metals, titanium, aluminum and silver wire embedded in a Geode.   This is followed by a wide assortment of electrical and battery devices, what appear to be light bulbs, and hundreds of types of aircraft. Besides all this technology, they had very early on developed ways to modify DNA strings. We do it today and make fruit-flies with legs growing where eyes should be, but they had turned it into an art-form. The dinosaurs had all died off thousands of years before so the Anakim remade them during the Pleistocene and probably in the Holocene after the Pleistocene Extinction that shifted the Earth and flooded the entire world.   We know this because, researchers started finding un-fossilized, elastic, dinosaur remains. Some even had viable DNA segments.

Dinosaur Reconstruction

Today we have hundreds of recently-dead dinosaur remains being studied all over the world. The following is a tiny sampling of these dinosaur DNA samples that may one day produce the Jurassic Park animals of the movies.

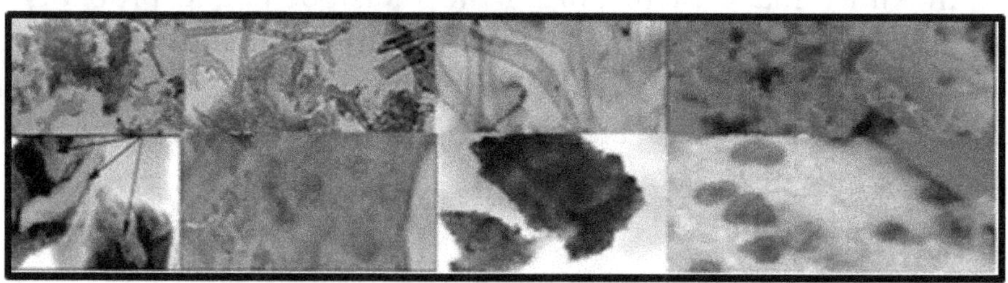

Probably everyone knows that God got mad at all the making of animals and messing with his special DNA. We are told God called most of the animals 'UNCLEAN' or worthless abominations, but the Anakim and now humans continued designing new ones so God destroyed everything again to slow them down.

Radiation-There were problems with dinosaur remains. Some were found with high levels of radiation which didn't fit the ancient Cretaceous destruction. T-Rex dinosaur bones had to be painted with lead paint or they would be dangerous for people to get near. How did they get nuclear radiations? Then something else started making us change our minds about massive beasts we though had been extinct. More and more are now claiming that dinosaurs were not running around even hundreds of thousands of years ago. Instead, <u>at least some</u> were only very recently living on earth and may have died just before the Pleistocene Extinction or even into the present Age. The evidence being found meant dinosaurs made it past the Cretaceous Extinction, or SOMEONE remade these dinosaurs from DNA samples from the Cretaceous monsters.

The list of dinosaurs that were, apparently remade during the Pleistocene keep growing and now include; Hadrosaur, titanosaurs, mosasaur, triceratops, ornithomimosaur, Lufengosaurs, T. Rex, and Archaeopteryx. Triceratops and Hadrosaur bones from Montana were tested for Carbon 14 two different dating labs both said that a triceratops registered an average of 31,000 years old and 23,000 years was the date for a Hadrosaur. The scientific community went berserk. This was a lie; this was a mistake; this was an anomaly. Soon all the back pointing was of no use as more and more finds showed the same thing. Some T-Rexes lived well past the time of the reported extinction.

Who Made New Dinosaurs?-Some will come up to you and claim the earth is only a couple thousand years old because some of the dinosaurs have been found that were not completely fossilized. There are countless writings on the regeneration of animals into something the Bible called "Unclean" or Abominable" animals. I have picked out a few to reinforce this substantial event that has greatly mystified geo-historians, religious historians, and just about everyone else. The details are clear as you will see. The Anakim scientists were making animals, but as we just read, Adam's children made "Unclean Animals" and dinosaurs as well. Here are some of the examples of what was said about the making of animals. Just about everywhere you look; we find that animals were genetically modified including Dinosaurs. Some may have even been regenerated after the worldwide flood.

Enoch 7:5-6 And the ANAKIM people began to <u>sin against birds, and beasts, and reptiles, and fish,</u>

*Jubilees 4:8- And lawlessness increased on the earth and **all flesh** <u>corrupted its way,</u> alike men and cattle and beasts and birds and after this* [A huge Pleistocene War] *they <u>sinned against the beasts and birds.</u>*

***Jasher 4:18-19** and the sons of men in those days took from the cattle of the Earth, the beasts of the field and the fowls of the air, and <u>taught the mixture of animals of one species with the other,</u>*

***Book of Naphtali 1:25-26-** The Gentiles <u>changed the order of nature.</u>*

***II Enoch 59:5-6-** <u>he who does any beast any injury in secret, it is evil practice,</u>*

***Book of Creation-** God was filled with anger and said "Samael, there is an immortal man who will appear <u>amid the creatures you have made,</u> ---Then <u>he and his followers made a great war</u>*

Way back in 1954 – Paleobiologists found soft tissue amino acids that supposedly lived during the Mesozoic. It was disregarded as impossible. After only 60 years most are realizing dinosaurs were here during the Pleistocene.

The Planet Venus would not make it out of the Pleistocene and almost instantly, the livable planet became lifeless and deadly to life. Its destruction caused a huge meteor showed that peppered the earth with millions of pieces of what is now believed to have been the Moon of Venus. <u>Over a million craters, now called Carolina Bays,</u> pot-marked the entire Eastern coastline of the United States and around to the land of Australia. The onslaught caused the Earth to destabilize and it shifted on its axis causing the Poles to melt and refreeze and the atmosphere to change completely as it rained for 40 days and almost everything was gone as mile high tidal-waves smashed over the tops of the tallest mountains.

Pleistocene Extinction Survivors

In Genesis 6 and 7 Moses stated 7-times that only people staying on the ground would be killed and other books like the book of Jubilees stated the same thing.

Genesis 6:16-7:23- *I will destroy everything <u>that is on the ground-- </u> And all flesh died that moved upon the ground; -- everything <u>on the dry land</u> died. -- Every living thing <u>upon the ground,</u> was blotted out. <u>Noah and those that were with him in the ark were saved.</u>*

Jubilees 20 *And God said that he would destroy everything which was <u>on the ground</u>, -all which <u>moves on the ground</u>.*

Surviving in Enclosed Ships

To survive, there were many ships described in ancient histories around the world. Of the three methods used, this one had to be the worst during the first 40 days as the Earth axis shifted and mile-high tidal-waves flipped ships over a number of times. In the book of Jasher we read it was like a 'kettle in a cauldron'. I don't know what that means, but we can imagine fear enveloped all and then the endless days waiting for the poles to refreeze and bring the water level down must have been nasty, but this was not the only way for survivors escape the flood.

Surviving in Flying fortresses

There were massive numbers of flying ships during this time, but the question might be how did they continue to fly? The answers might be provided in Sumerian and Jewish histories.

Enoch[19] ***74:15-*** *Enoch saw the chariots in the heavens. They were running in the world above to the gates in which the stars turn but*

---

*19 **Enoch-** Much older than Genesis, it was a source book transcribed from* *which never set. He also indicated that one of the "chariots of the heavens" was greater than all the other flying machines that travel around the world.*

Sumerian "Tablets of Enki"[20]- *An ark was built and the <u>seed of every animal</u> was transported along with the human survivors. <u>Above the Earth [in the Enoch-type-space-station], the Anunnaki cried out in fear.</u> Those saved from a worldwide flood sent out birds. God also sent a rainbow afterward. <u>The Anunnaki also survived.</u>*

Hindu Jain Texts-*"The Pushpaka Vimana was a gigantic 'Flying fortress the size of a large city entirely capable of holding unlimited numbers of people…Three flying-cities were made for and were used by the 'Demons'…One was in a stationary orbit in the sky, another moving in the sky and one was permanently stationed on the ground.*

Lhasa Sanskrit[21]*-They provided a description of how to build "Astras" (flying ships) including methods of propulsion, anti-gravity (laghima-a centrifugal force able to counteract gravity)*

Vaimanika Shastra[22]*-<u>Vimanas</u> were common during the ancient times and kept in a Vimana Griha, a kind of hanger, and were*

---

verbal histories. Found in several Bibles, this book was so revered as sacred scripture the Baptizing Essene made at least 13 copies of the book and used them often. During the time of Jesus, this text would have been well used by the Disciples and the Book of Jude has substantial references to details form this book.

20 Tablets of Enki- *These 14 cuneiform tablets described the beginning of the world just like. The big difference from Judeo-Christian texts is the Anak were called Anunnaki and Heaven was called Niberu.*

21 Lhasa Sanskrit- *These Chinese writings found in Lhasa Tibet. The writer is not known but it is of extreme age describing flying machines and powerful weapons in intimate detail.*

22 Vaimanika Shastra- *Transcribed around 450BC from very ancient works, this text by Bharadvajy the Wise provided intimate details about flying*

Egyptian-Hieroglyphics- *Osiris and Isis [Anak] descended from
the sky in a <u>sun-ship</u>, bringing wheat, and the arts of civilization to
the world.*

Here are a few of the depictions from India alone.

Indian Borobudur Temple- This single sight holds about 1500
flying saucer shaped stupas [Flying Machine effigies with hidden
pilots]. With the top section containing portals removed from one
of the effigies, we find an Indian performing what looks like the
duties of a helmsman [see middle image]. The main portion of the
temple holds a collection of hundreds of these single-man "fighter-
jets" of sorts to make the large "temple" appear to be a massive
flying aircraft carrier depiction from the time of the Bharata War.

***Drona Parva*[23]**- *In this ancient text, one Vimana described was shaped like a sphere and born along at great speed on a mighty wind generated by mercury.*

Surviving in Space

To make this even more bizarre, an ancient sculpting of one of the Sumerian transports used looks a whole lot like our space shuttles, with the same triple rocket design. OK! They only showed one astronaut in a full body suit, but you know what I mean. These guys had some sophisticated technology.

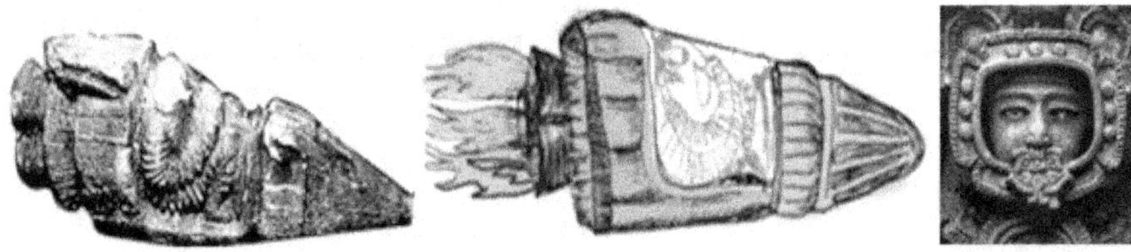

In India we find an ancient image of a helmeted man with some type of breathing component over his mouth, above right. Let's say not everyone left for the massive space stations that must have lined the Earth. Some may even have gone to various colonies on nearby planetoids of Mars and the Moon.

***Chilam Balam*[24]** *-Beings arrived from the sky on flying ships- white gods who <u>fly above the spheres</u> and reach the stars."- "Behold the circular chariot of gold, measuring 12 thousand cubits in circumference and able to reach the stars."*

Surviving Underwater

While we don't know if massive underwater Vailixi could have been much better than the surface ships, I'm sure some used the

---

*23 **Drona Parva**- part of the great Indian Masterpieces of the ancient times, Mahabarata, and the Ramayana.*

*24 **Chilam Balam**- One of the most sacred books of the Maya. Written in Maya by their greatest prophet Chilam, it was last transcribed around 1700 using ancient writings and verbal stories.*

underwater transports described in Jewish and Indian histories describe underwater

Hindu Creation Folklore--*In Indian folklore, Asvins [flying ships] from Atlantis were more advanced than vimanas [from India] and their Vailixi were cigar shaped and could go <u>in air and underwater</u>.*

Generations of Adam 6:1-4 *Among our little ones was Timnor. Timnor understood physical law and created mighty machines. Yea, Timnor built great engines with which to deface his Mother Earth, defiling her for his unholy purposes. With other machines of his contriving, his people <u>flew through the air</u> like birds and <u>explored the depths of the</u> rivers.*

Not only were these early people flying, but also they were using submarines according to this passage and the Hindu references.

Surviving in sealed underground Cities

Rebuilding ancient American locations and others across the globe may have taken decades. And the memory or threat of a new war got the Mongulala and others ruled by the Giant overlords who survived the Pleistocene Extinction. The Judeo-Christian book "Enoch" indicated a massive orbiting city circumnavigated the skies, and the ancient Chaldean "Sifrala" described how to make and operated a flying city. Babylonian "Hakatha" indicated the flying machines saved lives. Let's read a few excerpts.

Generations of Adam 6:4-5- *Timnor [a grandson of Adam] built great machines. With them, his people flew through the air and explored the watery depths. He created also great instruments of destruction.*

Survival Underground

Possibly the most reasonable if you can keep it sealed was what we find in the Persian historical records. We find something similar in the texts of the Mongulala. Just like many of the other people from before the flood, underground cities were required and lived in. This is the Noah story with a twist. The Aryan people were at a location that froze from the shift rather than being flooded and instead of riding out the storm in a boat, this massive Vara became sort of a sealed underground city.

Venidad[25]-Make a mighty VARA, an enclosure as long as a riding ground, with equal four sides. Here bring the families of Aryan men and women, cattle, dogs, birds and the red-flaming-fire. - Those that had lived in the bosom of the dale shall take shelter in the underground abodes.--There he brought the seeds of every kind of animal, every kind of tree, every kind of fruit. All those seeds he brought, two. ---In the largest part of the vara he made nine streets. To the streets he brought a thousand seeds of men and women. Inside the Vara, make water flow in a canal, one Hathra long. Keep earth inside the Vara, to grow green vegetables as food. Make cattle pens, to house the cattle of the Aryan people. Make three levels in the mighty Vara. In the uppermost level, make nine such wide enclosures, in the middle level, make six. In the lowermost level, make three enclosures. In the uppermost wide enclosure let 1000 couples, men and women live. In the middle let 600 couples live, in the lowest enclosure let 300 couples pass their lives. By self-light should the inner parts be illumined. The people in the Vara cannot see the stars, since the Vara is underground. That is why the Vara has lights of its own.

By self-light the inner parts were illumined because this massive survival module might have simply been underground. I told you it was like a city.-This artificial lighting is an example of the technology from before the flood. In South America we find a somewhat similar survival tool.

---

25 Venidad- This Zoroastrian/ Persian holy book is believed to have been collected around 1000BC and describes how the Aryans survived the flood by going underground in a sealed city or submarine.

Mongulala Description-*There are also underground cities inside the mountains we call the Andes. Tunnels link "lower Akakor" with the underground cities. These tunnels are large enough to accommodate for five men walking upright and they are so extensive that it takes many days to reach one of the other cities. The underground dwellings were the most wonderful of all. And they were given to the Mongulala. Great boring tools bore <u>through mountains.</u> It had a diameter of about 12 feet but only could hold 2. The designed boring machine for building Tunnel Cities in Akakor. There are also underground cities inside the mountains we call the Andes. Tunnels link "lower Akakor" with the underground cities. These tunnels are large enough to accommodate for five men walking upright and they are so extensive that it takes many days to reach one of the other cities.*

Possibly, this seal boring system might have been able to build sealed underground areas that could withstand the horrors of the Earth shift and flood. No matter what; during the Pleistocene, there were only 10 viable human races living in Europe, the Middle East, Spain, Asia, Java, and Africa, as shown, who survived the Pleistocene Extinction, 8000BC. All were Cro-Magnon, Cro-Magnon Hybrids and Anakim as shown. These may or may not represent what these survivors looked like as they have this mutation, but people with only this one mutation are not alive today.

We are told about 10-percent of the Anakim who survived the Pleistocene War also survived the Pleistocene Extinction in massive flying fortresses that orbited the Earth. We know about Noah's Ark and we have found it along with 2 others that are petrified as well; one on Mount Ararat in Turkey; one on Mt. Lucie along the border of Iran and Turkey, and one near the Euphrates River. We also know that Noah and his son Shem began the Shemites as pure Cro-Magnons [Haplotype F] while one son, Japheth had 2 of his sons marry Anakim wives.

One named Targ became the matriarch of the Scythians [possibly Haplotype [I], and another married Gomer to begin the Greek [G]. Another of Noah's sons married various Anakim. The most famous was Nimrod who married Semiramis establishing the beginnings of the Canaanites. Another went on to East Africa to form the Hamites [J]. One we read about in Irish histories was Noah's daughter, Cassair and her Hybrid husband, Bitheus, who went north toward Ireland. We can see that a number of the ancient bloodlines A, B, C, and D all survived, or the Anakim had created hybrids with those base-gene makeups and they survived. We will never know exactly what happened, but we do know that within a couple thousand years and everything was back running as it had been before the flood. Huge cities arose and technology abounded. People were flying from one place to another. Trade between kingdoms was flourishing. Technology was advancing. We don't know how great the societies were because of a horrible war that would overtake the world, but we do get a sense, when reading histories found in Jewish scriptures, histories of the Ughu Mongulala or South America and the Pre-Maya of Central America.

Technology Before the War

We see no mutations of human DNA or changes after the horrors of the Pleistocene Extinction until about 5500 years ago. At that date everything went crazy with respect to DNA; with respect to the world civilizations; with respect to technology; and with respect to the entire world. What is so weird is ~~almost~~ no history textbooks describe this most critical time when our DNA were so changed and the entire world was changed forever. Jews called it the time "the world was divided", Mongulala called it the "Blood Age"; the Indians called it "the conversion to the Age of Kali"; and the Egyptians called "the cause of Zep-Tepi" [A New Beginning]. The Pre Maya ended their first 5000 year calendar and started a new one at the end of this great strife. It boggles my mind that the horrors of the first and worst world-war of the Holocene Age is not studied in our schools and not even mentioned. Instead, it is almost completely ignore even when it was the most important event to our DNA of everything combined. Before the huge onslaught of worldwide war, we, as a people, had a very high level of technology and civilization. After the wars, we find massive mutation, loss of physical and mental ability, the reduction of society to a Bronze Age existence, a huge reduction in lifetimes from over a thousand years to only 100 years. Cities were gone, radioactive remains of victims were left behind, city walls were melted, millions were dead and some of the most horrendous mutations of people were being born to survivors around the world. Many of those who survived, did it by building complex underground cities to protect the citizens from many horrors. We don't get too much information from the Jewish historians, but here is a little bit.

Jewish Description

***Jubilees 10: 24-26**-And he confounded their language, and they no longer understood one another's speech, and they ceased then to build the city and the tower. For this reason the whole land of Shinar is called Babel, because the Lord did there <u>confound all the language of the children of men</u>, and from thence they were dispersed into <u>their cities, each according to his language and his nation.</u>*

By saying it three times the writer is pretty sure, some limitation in man's capabilities occurred 5000 years ago and it happened to everyone on the earth, not just the tower builders. We can tell from this and other Hebrew texts, People could understand just about any language that was spoken understood the mechanics of and built mighty cities of stone a huge tower about ½ mile high, before a war would take it all away. The PreMaya and the Mongulala were more straight-forward as they were disgusted with themselves for losing so much.

PreMaya Description

***Popul Vuh**[26]- They [the original Cro Magnon descendants from Aztlan], <u>had the power of understanding</u>; they saw and could <u>immediately see far</u> [really good vision.] They succeeded in knowing everything that could be seen or known in the world. Things that were hidden in the distance they could see without moving first. [Out-of-body movement or television?]Their wisdom was great; they controlled the forests, the rocks, the lakes, the seas and the valleys. [Control over nature] They investigated the four corners of the earth, [Fast Transport] they investigated the four corners of heaven. [Space-travel] They investigated the round surface of the earth. [Circumnavigated the globe] Then one day the heart of heaven blew fog in their eyes. They could not see*

***26 Popul Vuh**- was the PreMaya Bible. Transcribed from the K'iche Indian descriptions of Guatemala. Translated in 1701, it covers the very ancient history of the PreMaya people* *clearly any more, like breathing on a mirror. Their eyes were covered and they could only see things that were nearby. This was the way that the wisdom and knowledge of these first people was destroyed.*

South American Description

Uga Mongulala Secret History[27]*-The gods taught us the secret of man, animals, and plants. [Anakim people teaching the Cro-Magnon Descendants]-The Blood Age was the beginning of the Mongulala history. It started immediately after the Golden Age, about 10,500BC. There was an island in the west and a gigantic mass of land [Cultural Island] in the northern part of the ocean. Both lands were buried under an enormous tidal wave during the first Great Catastrophe. It occurred towards the end of the war between the two divine races. The war between the two divine races did not only lay waste to the earth, but also to the worlds of Mars and Venus. [Space Travel and colonization] The secret documents are kept in the underground Great Temple of the Sun. The documents testify to the 11,000 years of history and are kept in a room which is hewn out of the rock. They tell about the matter from which everything is made. They tell about the course of the stars and the relationships in nature. The ancient priests explored the spiritual forces of man. Our priests [Cro-Magnon Descendants] have learned how to make objects fly through space. They learned how to open the body of the sick without touching it. They had ships faster than birds' flight, ships that reached their*

---

27 Ughu Mongulala History- Transcribed from verbal testimony of the last Prince of the Mongulala, these details were incorporated into a book in 1978. Since thin, many aspects of the description have been verified, including massive tunnel, underground villages, transport roadway across the Andes, the Blood Age War, Venus and Mars colonization, and many other elements.

goal without sails or oars. They know how to <u>transfer thought without words</u> over the greatest distances. By magic, the ancient ones <u>suspended the heaviest stones, flung lighting and melted rocks</u>.

Holocene Dinosaurs

As amazing as the soft tissue dinosaur remains of the Pleistocene is; we can be pretty sure some of the huge monsters survived in test-tubes as DNA. They were recreated, or remanufactured after the massive shift in the earth that ended the Pleistocene Age and flooded the world. You can't keep a good geneticist down and the scientists of the day began recreating dinosaurs that were smaller and more tolerant versions. Most of the newer dinosaurs were slightly smaller, but generally they were still quite scary and those in the oceans were still pretty huge. While other things were going on as I explained earlier, some of the animals miraculously survived the flood and, of course, some of the ancient humans survived. What we find is that earth could no longer support the really massive land giants. Substantial evidence shows there were dinosaurs here after the Pleistocene Extinction 10 thousand years ago. Two of these animals are listed in Jewish histories. These were the Leviathan and Behemoth. The Behemoth was the land variant.

Behemoth- In the centuries that followed the Cretaceous Extinction, some dinosaurs and other large creatures were regenerated. In Mexico and in Peru images of men with dinosaurs were drawn. As shown on the left below, in Peru, images of dinosaurs were often put on clay while clay models of dinosaurs and dragon-like creatures were molded in ancient Mexico, as shown on the right.

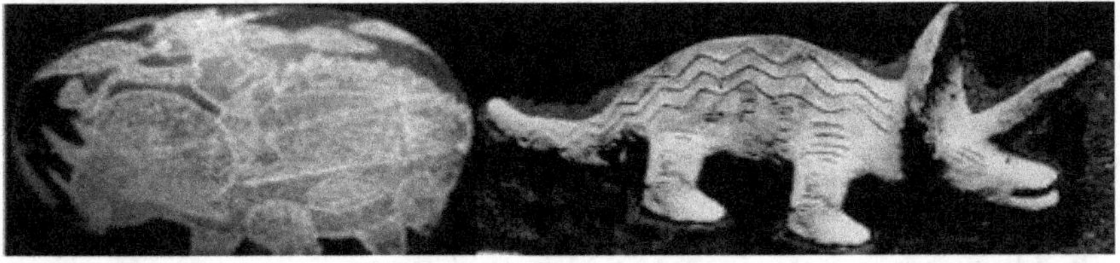

The images and stories indicate that, not only did the Earth inhabitants walk with dinosaurs during the Cretaceous Age, but humans were in company of dinosaurs in the Pleistocene and beyond. Below are some more of the pictures found in Peru which show that dinosaurs were still alive during much of the post-flood Era. The one on the left was a depiction found in a cave while the depiction on the right was a huge depiction inscribed in the Nazca Plain of Peru. One thing to note about the depiction on the Nazca Plain is that the drawing had to be done after the worldwide flood 10 thousand years ago. If it was drawn before the flood, it would have completely washed away.

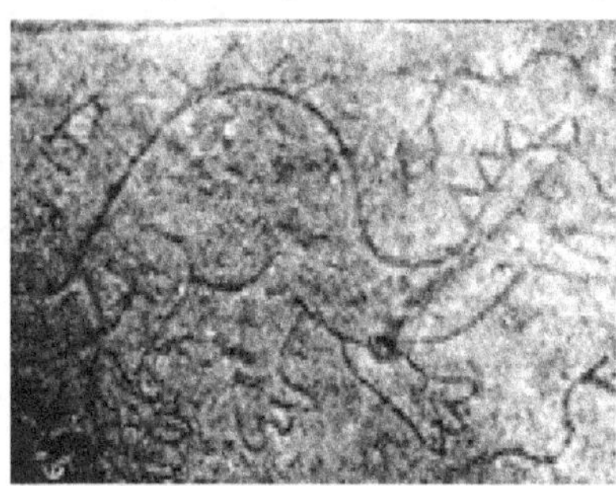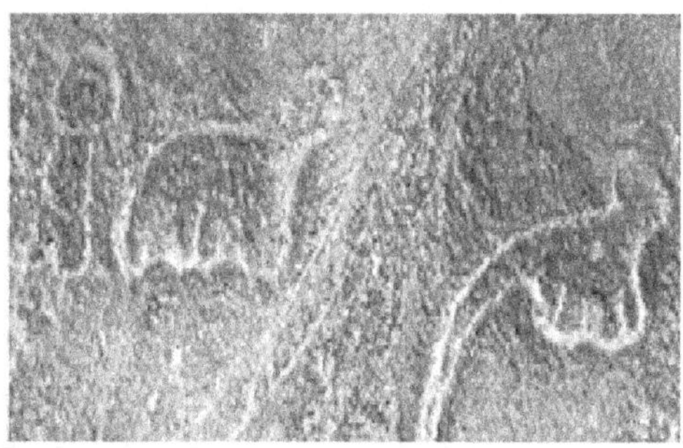

What these images and the hundreds of other similar ones show is not a mystery or an anomaly, it simply shows that genetics was mastered by people during the Pleistocene and probably by those who survived the Pleistocene Extinction.

Man almost certainly saw dinosaurs well after the dinosaurs were supposedly extinct at the end of the Cretaceous and that means there is a likelihood that they were not killed in the extinction or DNA was found and the dinosaurs were replicated.

It should be noted that even the Bible recounted dinosaurs surviving even well past the Pleistocene Extinction as the Jewish prophet Daniel killed one while serving the King of Persia, That means they survived the dinosaur extinction, and the Pleistocene Extinction. On land another dinosaur survivor was certainly noticed all over the world. At that time it was called Dragon or Behemoth. Here are a couple more Biblical texts.

II Esdras 6:48-Then he set apart two creatures the Behemoth and the Leviathan. You put them in separate places. The country of 1000 hills was given to behemoth. [Dinosaurs were remade.]

Job 40:15-24- Behold now behemoth, which I made with thee; he eateth grass as an ox. His strength is in his loins, and his force is in the navel of his belly. He moveth his tail like a cedar [Huge Diplodocus dinosaurs had a tail worthy of mentioning]: The sinews of his stones are wrapped together. His bones are as strong pieces of brass; his bones are like bars of iron. He is the chief of the ways of God: He lieth under the shady trees, in the cover of the reed, and ferns. Behold, he drinketh up a river, and hasteth not: He taketh it with his eyes: his nose pierceth through snares.

Palestrina Dinosaur-This mosaic done around 100AD shows a very impressive dinosaur being attacked. Maybe they were hunted like other animals a couple of thousand years ago. [Below left]

Greek Dinosaur-The image above right clearly shows people fighting an ancient behemoth. Most of the people of that time had never read about dinosaurs so their image of these creatures came from actual encounters. In this case, looks like the dinosaur is losing. The ANAKIM had created or recreated these animals during their reign on earth. Their reign lasted thousands of years with the help of something that kept them alive for thousands of years. Greeks called it Ambrosia, and the Jews thought that it was similar to pomegranate. Possibly it was simply mutation of human DNA that shortened life after 6000 years ago.

Anasazi Dinosaur-A very clear petroglyph of a dinosaur found at Natural Bridges National Monument in Utah. It is attributed to the Anasazi Indians who lived in the area about AD 400. The image below left shows a diplodocus like monster with a human nearby.

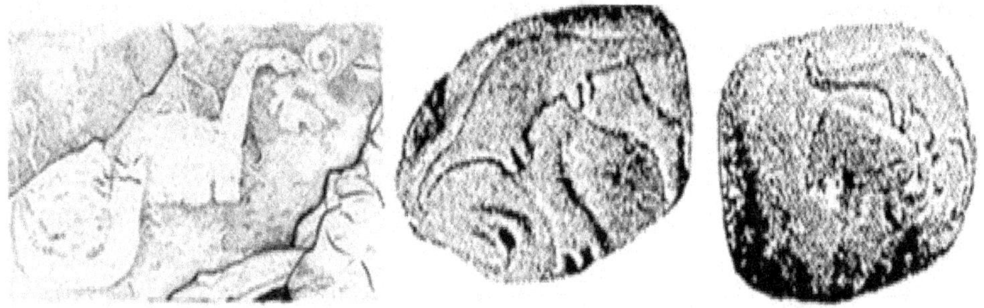

Colorado Monster- Huge granite carvings of the time show dinosaur-like monsters still roaming around. Quickly sketched so as not to be eaten, it looks like the thing was larger than the rhinoceros also depicted. [See above right]

Mayan Monsters- There can be little doubt that the ICA stone engravings show dinosaur-like monsters that roamed South and Central America in the not too distant past. [See following left].

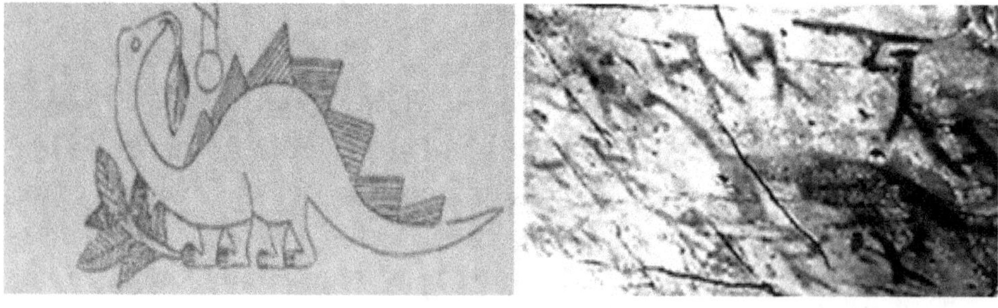

Peruvian Behemoth- On a rock cliff in Peru we find that these monsters were hunted by the ancient pre-Inca. [Above right]

Egyptian Dinosaurs-Possibly everyone has seen the dinosaurs from the tablet of Narmer, the first unifying King of Egypt. If not, it is shown next left.

Sumerian Duplicate-Some may not know how very close Sumeria and Egypt was during the first part of the Holocene Age, but, the dinosaur images above right are from Sumeria and an almost identical match even to the crossing of the necks. The question might be what places did not see these things? The reason I brought up these genetically modified animals is to give you a better image of what people were like during and after the Pleistocene; including the lineage that would carry God incarnate 20 thousand years later.

Worldwide War

Around 3500BC wars around the world broke out. Soon everyone was fighting everyone. If you remember the 16 Uranium processing sites[28] in Gibon, Africa; we are told enough processed Uranium has been taken to power the State of New York for a year. IF it wasn't nasty enough, if find that a massive 500 mile in diameter blast area in the Libyan Desert has turned the desert floor to glass. This is thousands of times as large as similar glass producing blast craters of today's nuclear weapons, so you can imagine how devastating the wars were. Nuclear World War abounded. Fortress walls around the world are being found melted, remanufactured dinosaur bones are so radioactive, they must be painted with heavy lead pain and remains of victim in what is now called the "Mound of the Dead [Mohen-jo-Daro] City are said to be still radioactive and all the clay pots in what remains of that city The Biblical book of Jasher tells us that 1/3 of the population of the entire world was destroyed in the war and 1/3 were so mutated that they became like animals. The remnants of bodies, still radioactive, have been found along with globs of glass that once were pottery. Melted stones on building and walls help us understand that nuclear bombardment also would cause substantial DNA Mutations. As the Wars subsided, around the world this time was proclaimed as the "New Beginning" "Zep-Tepi to the Egyptians, The new Age of Kali to the Hindu. The Pre-Maya started a brand new 5-thousand year calendar. On and on we could go. As the radiation began to subside, mutations of human DNA began to explode. Let's look how the mutations changed how long people lived.

---

28 Scientific America- *"Nature's Nuclear Reactors, the 2 billion Year old Fission reactors in Gibon" 2011- and many other articles describe the loss of substantial amounts of processed uranium with no explanation. Other details are found at IFLSience.com, Ancient-code.com and many others.*

Outcome of the War

Let me give you some figures to think about. The <u>human genetic diversity today is vastly different from what 200 generations ago</u>. A study dating the age of more than 1-million <u>single-letter mutations</u> in the human DNA code reveals that <u>most of these mutations are of recent origin</u>. About 90 <u>percent of the harmful single nucleotide mutations arose between 5 and 11 thousand years ago</u>. Oddly, since then there have been few mutations at all. Overall, researchers now believe that <u>about 81 percent of the single-nucleotide variants in the European sampled and 58 percent in the African DNA sampled arose in the past 5,000 years</u>.

Absolutely, the mutations could have been the result of some type of chemical or biological agent that spread during the war or there could have been a disruption of our atmosphere for a time that allowed massive amounts of cosmic energy to invade our DNA but the main thing is our DNA was changed. Let me quickly go over some evidence that helps explain this massive time of Babel War Mutation along with the Pleistocene Extension Mutations.

Many fortress walls around the world were melted during this war in many places. [Super-heated nuclear blasts?]

Large groups of human remains are piled up in the streets. [Victims had no way to get away just like Nagasaki]

Some human remains are even still slightly radioactive after over 5 thousand years.

Sixteen Mesozoic Nuclear processing areas found in Oklo mountain site of Gibon, Africa are missing enough processed Uranium to power the State of New York for over a year. These losses appear to have happened at least 5 thousand years ago.

During the Young Dryas [11 thousand years ago, huge spike of radioactivity and other signs of nuclear event was recorded. I showed the chart earlier.

Many unfossilized Tyrannosaurs Rex bones not only show they lived less than 20 thousand years ago but also they are so radioactive, they must be painted with lead based paint to protect viewers. This type of radiation could, certainly affect our DNA.

The City of Kukkutarma- Now called Mohen jo Daro is littered not only with bodies but also lumps of glass that used to be clay pots.

Puma-Punku Leveled -While most know about the Tower of Babel being flattened, on the other side of the world a massive temple complex in Peru called Puma-Punku was completely destroyed.   Some of the stones used were 26 feet long and 16 feet wide which should bring to mind the huge boulders at the Baalbek, Lebanon site.

Around the world we are finding massive colonies of people who moved to underground-cities 5-thousand-years-ago to protect themselves for something horrible on the surface. While going underground would not protect someone form conventional warfare, it could protect inhabitants form nuclear radiation. Many underground cities were found in Turkey, Malta, China, Scotland, Mexico, Peru, the U.S.A, and just about everywhere else.

Smaller Brain- After the War, our Brain began to atrophy from disuse from 1600cm³ of the Cro-Magnon to the 1350cm³ of "modern Man".

Jasher 7:19-20- the first son of Eber was Peleg-[Peleg means World-Divided] and the name of the second son was Yoktan, meaning that lifetimes of men were diminished. -God said let us confuse their tongues---And from that day following, the Lord confounded the Language of the whole earth;

These two mutations changed DNA as the brain atrophied.

Erra Epic[29]- The poison, by itself, it destroyed the city."-- "with the weapon that smites, he threw fire upon the mountains and smote the people."-- "The weapon robs the senses. --In a cavity inside the earth they dwell while the weapon destroyed the cities, Many hundreds of thousands of people died in the war-- A great land-annihilating storm...An evil wind clutched them and would not grant them another day... The face was made pale by the Evil Wind. It caused cities to be desolated.

Exploits of Ninurta[30]- A terrible weapon was used and launched into the enemy. It consumed animals in the open country like firewood. It roasted the animals like locusts. With the weapon he crushed their cities. People gasped for breath. The people that were not killed became ill. They hugged themselves. It rained down coals and flaming fires. The fire consumed men and reduced forests to heaps. The fields became like soot and the whole horizon was reddish like purple dye. The missile's venom alone destroyed townspeople.

Oera Linda[31] -In this year 3449, the air was heavy and oppressive. In the midst of this stillness the Earth began to tremble and

---

*29 **Erra Epic** –Babylonian Description of Nuclear Fallout during the Bharata wars*
*30 **Exploits of Ninurta**- Chaldean Description of Nuclear Fallout similar to Erra Epic.*

swallowed up many. Whole <u>forests were burned one after the other</u>, and our <u>land was covered with ashes</u>.

Marusala Parva[32]*-The Iron Thunderbolt weapon" [nuclear weapon] provided the power of the universe with incandescent column of smoke and fire, as brilliant as a thousand suns. Survivors had hair and nails fall out; birds turned white; and the only escape was under water. It reduced to ashes the entire race of Vrshnis and Andhakas. Pottery broke without cause and the food was infected.*

Mahabharata[33]*-Depictions of Mushroom clouds from major blasts with Elephants turning white, Animals turning to dust, people running into the water for relief, and thousands dying all sound like nuclear bombs.*

Rig Veda[34]*-The <u>Thunder missile</u>, of the Aryan's might and glory. A massive explosion demolished Sambara's hundred ancient walls; who cast down Varcin's sons, a hundred thousand. Thy <u>Thunder missile</u>, at the Dasyu; increase the Aryan's might and*

---

*31 **Oera Linda**- Celtic manuscript written in Old Frisian around 1000AD, to cover historical and religious themes of remote antiquity, from 2194 BC to 800.*

*32 **Marusala Parva**- [Book of Clubs] is one of the 18 books of the Mahabharata. It describes the destruction of the Yadavas through in the 36 years of the Bharata War.*

*33 **Mahabharata**- traditionally attributed to Vyasa around 400BC, this Sanskrit text is one of the two major Sanskirt epics of ancient India, the other being the Rāmāyana. It narrates the struggle between two groups of cousins in the Kurukshetra/ Bharata/Tower of Babel War.*

*34 **Rig Veda** is an ancient Indian collection of Vedic Sanskrit hymns. The Rigveda is the oldest known Vedic Sanskrit text and one of the oldest extant texts in any Indo-European language but orally transmitted since the 2nd millennium BC.*

Massive Mutation Described-*Jasher 9:35* *A third of all the workers became like apes and elephants.*

Ramayana[35] ***I 17:8-18- *Vanaras [ape-people] are powerful as the gods began to parent sons in the semblance of monkeys.***

Lower Congo Ape men-*"First God created man. After a huge flood, men put their milk stick behind them and were turned into monkeys."*

Mayan Tradition-*During the second creation, people turned into monkeys and the world was destroyed by wind.* Possibly talking about the destruction of Babel and another indication of people turning into monkeys.

Aztec History- *according to Codex "Laticano-Vatino"-During the age of the four winds men turned into monkeys.* Possibly, wind destroyed the tower and the <u>monkey</u> thing keeps coming back.

Called the Bharata War by in the Indian histories, this time of massive mutation of all the people of the world was a horrible time not brought out in ~~many~~ any classrooms. Some of the mutations were noted by major changed in how long people lived or how they could use their massive brain, but the worst mutation was more than simply establishing a new race of people called the Vanara.

People were living like animals or being animals, the Anak rulers were either extremely old or dead and the mutations quickly changed how the Americans could think. For a couple thousand years, everyone on the planet had to start over. It seems that during this transition time, 4 groups reestablished a life in the Americas as the Adena in North America began to recover, the Mongulala in Eastern South America, the preInca in Western-South America,

---

*35 **Ramayana**- [Rama's Journey]-This great Sanskrit poetic history of the Dravidians in India during the Bharata War was transcribed about 600BC by Valimki and is actually made up of 7 books.*

and a 4th group made up the preMaya/Aztec. It seemed these people all had Anak giant overlords. Some were benevolent, and others were not so much. Let's look at a timeline now that you have more insight.

Let me ask you a serious question; *"How could almost the identical stories be found around the world?"* Assume for a minute that hundreds of ancient people were not completely wrong in their observations. All we can assume at this time is that a large group of humans had mutated by chemical, biological, or nuclear fallout means. Possibly, they devolved back to something similar to Homo-Neanderthalis, but these unfortunate people became useful in the societies of that time and many were well respected. From the information in the book of Jasher, we can believe as many as 1/3 of the population of the world was mutated in this or a similar way. A number of mutations, such as donkey and burro that cannot procreate easily and would have died off if it weren't for the back breeding with horses. Possibly, that was the Vanara's fate as their populations dwindled over time. After about 500 years, we find only tiny groups of hidden away groups. Between 3000 and 2500 BC, ape statues Egypt, China, India, Cambodia, West Africa, Central and South America all suggest that the monkey-gods were feared, adored, and worshipped, but most of all, they had integrated with societies. So what do we have?

- **Massive amounts of written details** from Inca, Aztec, Indian, Japanese, concerning the mutation of a large segment of people to become ape-men and many became great leaders just after the war.
- **Massive numbers of ape-men depictions** in artwork, etchings, and statures showing the ape-like resemblance from Western Africa, Egypt, Cambodia, India, China, and Central America.
- **Written and oral histories** of a monkey man coming to Central America to attack the ruler of Teotihuacan also known

as Patala Loka from India and PreMaya writings and Akahim in Brazilian histories.

- **Similar simian characteristics** of Hanuman, Quetzalcoatl, Thoth, Pharaoh Narmer, Thoth's son, and the Japanese variant of the same history.
- **Indian historical data** describes the location of Patala Loka and being on the opposite side of the world from India placing it in Central America
- **Similar Homo-Cognatus** possibly living today helping to confirm the ancient mutation
- **A possible frozen specimen** of Homo-Cognatus was put on display to help confirm Vanara people.
- **Similarity of** Tlaloc and Ravana in Indian and Maya historical records.
- **Similarity of Quetzalcoatl** and Hanuman in Indian and Maya historical records.
- **Remains of a PreMaya city** worshipping a scary looking Ape-man.
- **Remains of a Hindu** Worship Center in Peru. How did it get there?
- **Ancient religious and historical documents** from Jews, Hindu, describing Flying machines and details of the ape-men Flying which allows for the trip around ½ way around the world to be practical.
- **Ancient flying machine artifacts** found all over the place.
- **Details of a high level of civilization** and technologic skills to build and use flying machines and weapons of war needed to cause massive mutations.
- **Knowledge that ½ of all human mutations** occurring when the Homo-Cognatus and Pan Troglodyte mutations occurred.
- **Knowledge of Indian people DNA** from 5500 years ago being found in Central America makes us think Indian people were here about that time.

- **Description in Egyptian historical records** of a large number of the Anak type god-men leaving Egypt to all parts of the Earth to set up new colonies before the Bharata War.

- **Descriptions of the horrors of the war**, nuclear fallout, having to live underground for a time, commonly showing comparative details around the world to help confirm the hi-tech war.

- **Neanderthal and Cro-Magnon brains** being so very much larger than ours showing why humans were thrust back into a Bronze Age around 3000 BC.

On and on we could go. While it is impossible to get a 100% assurance of a history. That being said, let me continue on as they one thing that is 100% known. 5500 years ago, something happened to our DNA.

Mutations After the War

After the worldwide war or 5500 BC, whole scale mutation of humans changed the entirety of humanity. These 14 human races along with the predecessor 12 human types would continue into the present age, but there were significant changes to the DNA.

The number of mutations had doubled almost overnight. The Homo-Capensis, long headed people had changed from huge giants to moderately tall, red-headed, Long-thin-skulled, rulers and Cro-Magnon hybrids had expanded to 25 or so races of humans who were interbreeding to make a more homogenous population. The African nations seemed to stay by themselves more so DNA was not mixed between the 4 or 5 African mutations and the Eurasian types and the brand new "Q" race dominated the Americas. These people somehow scattered across the globe.

Haplotracking Human-DNA

The following charts allow us to witness the modifications of the DNA of the Humans that were living so long ago. The reason it is centralized is because it only is tracking Cro-Magnon humans and early hybrid humans who carried Anakim DNA.

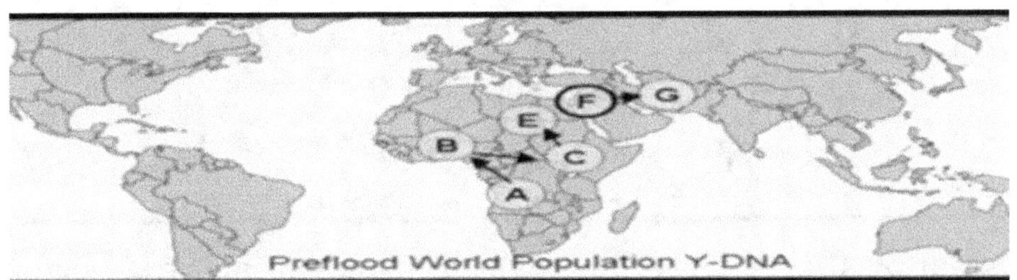

The following chart shows an explosion of DNA mutation happening in a very short time. The reason for the wide distribution is these sight were where people landed after the flood.

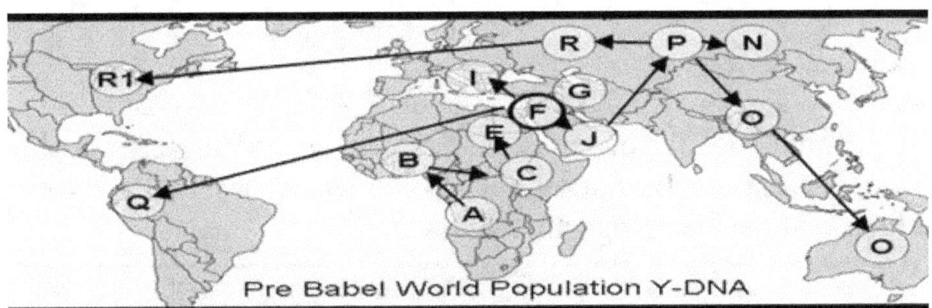

The last map show the second blast of mutation that occurred 5500 years ago and where populations were at that mutation time.

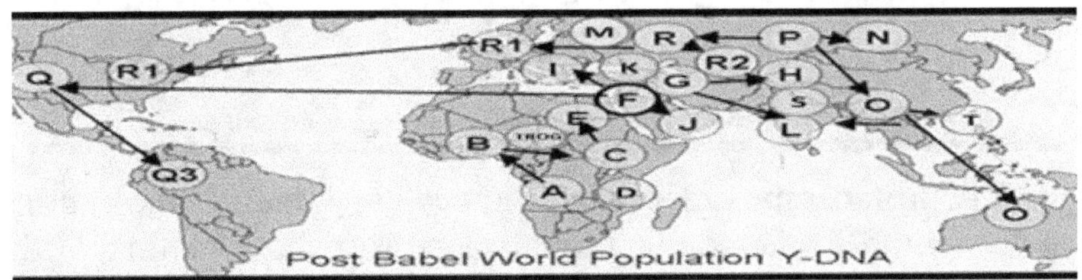

Haplotracking MtDNA

When scientists tracked Mitochondrial DNA, they found similar things. Some of the Anakim DNA that was used to produce the primate was implanted into Cro-Magnon to build hybrids. This was possibly done by sexual intercourse.

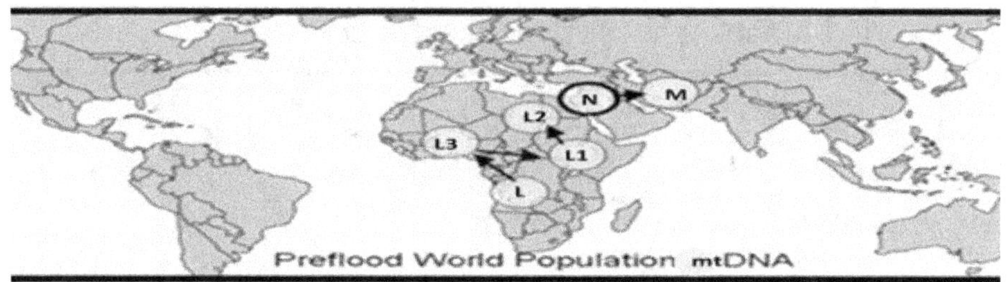

The arrows generally show which mutation occurred before another, but they all generally happened at the same time.

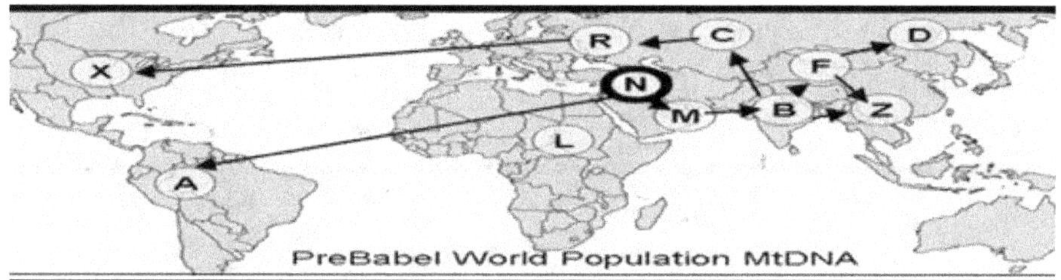

The peculiar transfer of DNA to the Americas 5500 years ago not only show massive mutation but also shows a mass transportation method that has been lost since the war.

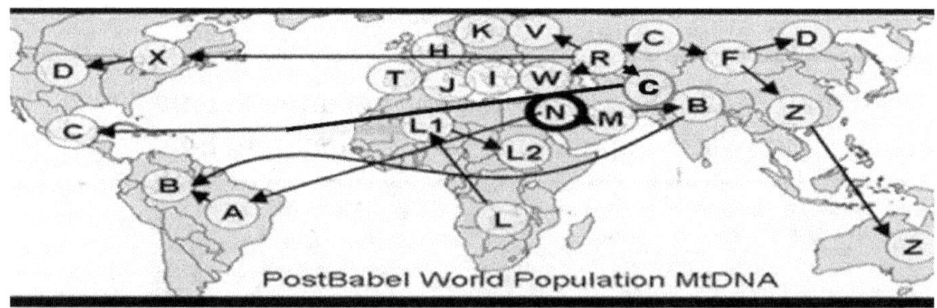

The Y-Chromosome mapping showed a strange and abrupt transfer of the F and R mutational races somehow getting to the

Americas at the beginning of the Holocene without going through Asia or anything. We see a similarity in the MtDNA track as the R and N variants of Mitochondria from the Scythians and middle Easterners appear in the Americas. This connection with India gives credence to the Stories of Hanuman leaving India to settle the Americas, but who knows? One thing we do know is that people used to live thousands of years. Adam, in the Bible lived 930 Jubilee years or 6500 solar years and Noah had a similar lifespan. We can track King Nimrod's life to be about 7000 years and King Thoth lived well over 10 thousands years, with a little help. On and on we can go and it always comes out the same but after the Bharata war, lifespans began to quickly shorten, Peleg, Reu, and Eber who all lived through the war only lived about 235 years., Abraham's dad only lived a little over 200 years and Abraham only lived 175 years. Without a doubt people do not live as long since the war.

Lifetime Mutation

As populations erupted all over the place, one thing that is very obvious. People no longer lived very many years. By simply looking to see how long people lived we can possibly determine a lot about the character of a timeline and how it should be modified to allow critical events to line up in tales around the world. The biggest indicator happened during what has been called the Tower of Babel War, simply because the tower was destroyed at that time. As part of the massive modification of DNA around the world, sometime during the end of that war, lifetimes got much, much shorter.

As the populations tried their best to save themselves from the nuclear radiation, many changes were happening to their DNA. One change is very obvious in all ancient histories as shown in the following graphs. I have timed the beginning to about 40 thousand years ago for Cro-Magnon, that would represent only 8 or 9 generations which is confirmed in a number of Jewish texts. If we read the ancient Jewish book "Adam and Eve II", it states Adam lived about 6500 years and we can assume the other Adamic people before the flood lived about the same length of time. The Genesis book described Adam living 930 Jubilee-years which agrees with other histories. The charts following are the Jewish and Sumerian king lines. These are followed by similar timelines of the Chaldeans and Egyptians. All are similar. The Chaldean and Egyptian timelines were compressed slightly to line up the critical events, but the slopes are the things to look at. Overnight, all societies around the world were affected by something and life-spans were truncated. Six thousand year lifespans turned into 150 years and less over one generation.

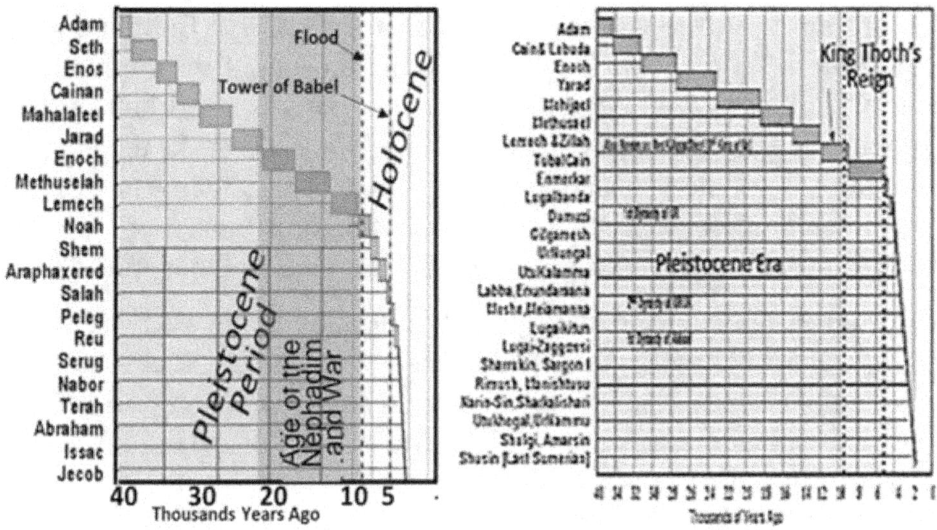

Lemech Marker-According to the book of "Jasher", Lemech, in the line of Adam, was ruling during the Venus destruction time so we can get a time marker from his reign itself by timing the massive meteorite damage known as the Carolina Bays that deposited over ½ million craters on the East Coast of the United States about 11 thousand years ago.

Noah Marker-Noah, of course, survived the worldwide flood. So there is certainly a marker with that Patriarch. Others that survived also can provide this same mark in time. Genesis tells us 4 times that all men <u>remaining on land</u> died during the flood so we know only those above the earth, under the sea or on boats survived. The Sumerian history tells us many of the Anakim people cried out in fear as they weathered the flood in flying ships.

Peleg Marker-Also it should be noted that Peleg ruled during the time of the great Tower of Babel War. His name means "<u>when the world was divided</u>". Therefore, his reign coincides with the 6 thousand year old time.

According to Hittite works *"The Epic of Creation"* (Enuma Elish) and *"Epic of Gilgamesh"*, the Sumerians tell about the same story as the Chaldeans and the Jewish histories. This is important in that Gilgamesh and Noah may have been the same King. In the *"Book of Giants"* (ancient Dead Sea Scroll book) Gilgamesh was a prominent authority figure who was "the only survivor" of the flood. In the Sumerian king line, it appears he was Uber-tutu.

The reason Chaldean and Egyptian timelines begin early is their rulers were Anak who ruled 100 thousand years ago. Please notice this same mutation noted after 6 thousand years ago. In every case, life spans were greatly reduced to about 150 years and even shorter soon after the war.

The Egyptian monarchs of the "0th dynasty" had the same unusual life-spans. All of a sudden, 6 thousand years ago, something happened and the ruling people began to get shorter lives. The major mutation events are depicted on each of the charts to show consistency and strangeness. The Egyptian timeline is shown next. King Thoth reportedly was in the massive war before the Egyptian "Zep-Tepi" (new beginning}.

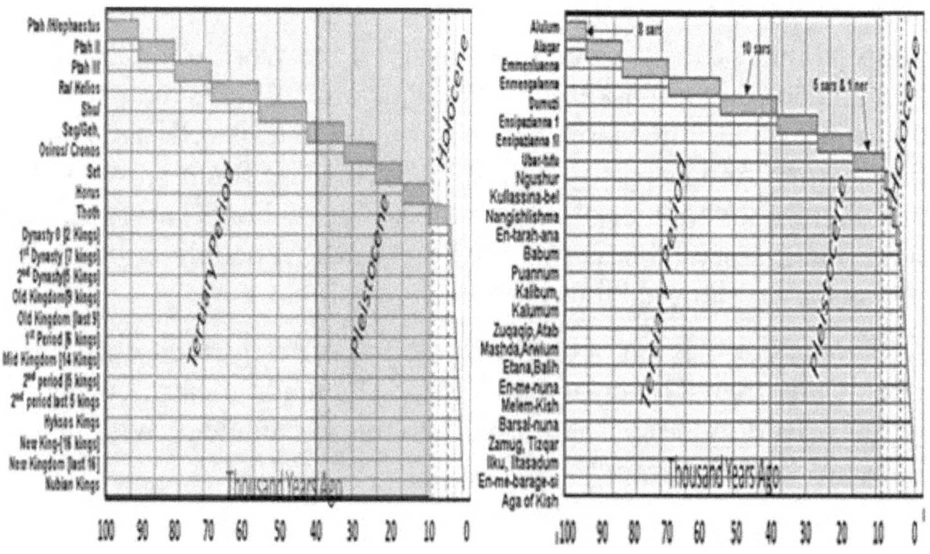

Cro-Magnon Marker- Timing analysis of bones, complex tools that were built, and village life noted points to a 30-40 thousand year old event. Suddenly, there were Cro-Magnon type humans. It should be noted that much to the timing was from nuclear decay so the 40 thousand years might need to be compressed a little. Knowing he was a separate creation from Neanderthal helps as we can quickly see that Cro-Magnon's emergence quickly changed everything about human existence. The characterization of the "new man" described in Genesis was so striking; we label it as the beginning of the Pleistocene.

Pyramid Marker-We can date the building of what is known as the great Pyramid of Giza to this same general time so it is reasonable to place the 1st recognized king of Egypt, Thoth, as starting his reign 15 thousand years ago.

Adam Historical Reference-The books *"Adam and Eve[36] I and II"* tell us that there was 5500-solar-years between the birth of Seth and Adam's death. With Seth being born about 130-jubilee-years or 1000 solar-years after Adam's creation, Adam lived 6500-solar-years or 930-jubilee-years. For cross comparison, Jubilee-years are described as normal timing in Daniel and in Deuteronomy. By this small revelation the Adamic Kings listed in Genesis before the flood, superimpose the dates and timelines directly over the physical evidence elements showing science, history, and religious documents very closely using the Pleistocene Extinction and Worldwide flood as Noah's flood.

Adam and Eve I 3:2--Yea the word that will again save thee [Adam] when the 5½ days are fulfilled – the God in his mercy explained to him that these were 5500 years.--21:9--"O Adam all

---

36 Adam and Eve I and II- Also called Apocryphon of Moses, these sacred books are found in a number modern Bibles including the Messianic Essene Bibles and 6 hand written copies were found among the Dead Sea Scrolls.

the misery will not alter the covenant of the 5500 years.--38:2- O Adam, when 5500 years are fulfilled, thou shalt live forever.

Genesis 5:5 *Adam lived 130 [Jubilee] years, and begot a son in his own likeness. --Adam lived a total of 930 [jubilee] years, and then he died.* 6500 Solar-Years

Adam and Eve II 19:1— *[sometime after Seth was born] God revealed to Adam again concerning the 5500 years [+910 solar years to the birth of Seth].* Adam's lifespan over 6410 solar years

Daniel Jubilee 9:27-28-*Seventy 'jubilee-years' [483-solar-years after 445-BC] are decreed for your people. There will be seven "jubilee-years" [49 solar-years] and sixty-two 'jubilee-years' [427-solar-years] After the sixty-two 'jubilee-years' [434 solar-years] the Anointed One will be put to death [Jesus would die 5-AD]* Interpreted in Daniel as "Seven" we know this was really a 7-year period that was used to track time.

Jubilees 4:1-11 *And in the 3rd <u>week Eve gave birth to Cain</u> in the 4th <u>week </u>she gave birth to Abel. In the 5th week she gave birth to Awan.* A week-of-years or Jubilee was what it was saying.

Deuteronomy [Jubilee] 15:1-*At the end of every seven- solar-years you must cancel debts.* This was a jubilee-year.

Deuteronomy [Jubilee] 31:10- *every seven-solar-years you shall make a release.* This was a jubilee-year.

Nehemiah [Jubilee] 10:31 -*Every seventh-solar-year we will let the fields lie fallow, and will cancel every debt.* This was a jubilee-year.

II Esdras[37] 14:10-12 Marker-*For the world hath lost his youth, and the times begin to wax old. For the world is divided into*

---

*37 -**II Esdras**- Also known as 5 Ezra transcribed around 90AD, this is from the King James Bible, is in many different Bibles, and is considered sacred and holy.2*

twelve parts, and the ten parts of it are gone already, and half of a tenth part: And there remaineth that which is after the half of the tenth part.

These writings are about Ezra who lived about 3500 years ago. If we assume there is about 1500 years to go before the Earth is destroyed so 3500 years is 3/24 the total time man has been around or 28 thousand years before the time of Esdras or 31 thousand years ago. Assuming that the world will have a terrible calamity, in less than 1500 years is not only a good guess, it has been prophesized by "seers" from around the world and throughout the ages.

Manetho History of Egypt—*[Greek History 200BC] Gods ruled for 13,777 years followed by 15,150 years of rule by demigods and spirits before the end of the last epoch.*

When added to the 10,000 years since the flood the last significant epoch, the first Anakim ruled Egypt <u>39 thousand years ago</u>,

Diogenes Laertius History- [Greece 340AD], *the astronomical records of the Egyptian priests began in <u>49,219 BC.</u>*

Turin Papyrus- *Gods ruled Egypt for 13,420 years followed by 23,200 years of rule by demigods*

When added to the 5 thousand years since the time it was written, the Anakim ruled about <u>41 thousand years ago</u>.]

Emerald Tablets-   According to this Egyptian text, originally written by Thoth himself, Thoth was sent to Egypt from Atlantis before it sank. *He was sent 37 thousand years ago; he became the ruler; and at some point, he created the first 2 pyramids.*

Assuming the 30 thousand years started near the time of Adam and Eve, and the reference is about 5 thousand years old, then Adam and Eve would have been here about <u>43 thousand years ago</u>.

Popul Vuh-[Ancient PreMaya writing] - *The age that followed the "age of primitive man" was the "Age of heroes* [Anakim-called

Titans] *and demigods". Sixteen sons of the Gods* [Anakim] *ruled the land followed by 45 priests* [Anakim hybrids]. *After this time, there was destruction by flood and 2 were saved. The survivor sent a falcon to test the land and finally God sent a rainbow.*

Although the time is not indicated to be the 40 thousand years, it is not hard to assume that with life spans of almost a thousand years that this too represents a 40 thousand year era.

Syncellus History- [Byzantine history 950AD] *The chroniclers of the pharaohs had recorded events for 36,525 years.*

Martianus Capella History- [Roman 540AD] - *Egyptian Sages had secretly studied astronomy for 40 thousand years before they imparted their knowledge to the world.*

Chaldean History- They indicated *there was 39 thousand years between the first dynasty and the flood.*

One thing that affected lifespans was marrying Anakim who lived longer. Cush, Noah's grandson married an Anakim as did Magog, another one of Noah's grandsons. Anyway; you look at it, DNA was not only modified to reduce brain function, but also to reduce lifespan. Even Zeus plays into this part of the story.

Here Comes Zeus

It is fairly easy to find out when Anakim bloodlines entered the otherwise Cro-Magnon Base after the Flood. Sometimes it seems fanciful because Anakim were described as gods in many societies. According to Irish historians, Japheth's son, Magog, was in the line of the Irish and Magog wandered from the ways of his father Noah. His wife was named Targ. Now Targ was either a full blooded Anakim or related in a strong way. Her Great, Great, grandfather was reported to be the famous Zeus [King of the Anakim kingdom of the Olympians near the Geek area settled by the descendants of Japheth. Please understand, what I'm saying is that the main block of Irish descendants were all Anakim-Cro-Magnon Hybrids. If you are wondering about the Anakim lineage, Zeus' son Targitaus and Borysethenes' daughter had a boy named Colaxias, who fathered Scohoti; who fathered Targ. The son of Magog and Targ was named Massagetae and he is credited as being the first of the Scythians and Samaritans who later settled near Greece. Hercules is somewhere in that whole mess, but I just want you to know after the flood, just about everyone was becoming or had already been an Anakim-Cro-Magnon Hybrid. Either Massagetae or his son, Scythes, were in control of the Scythians whenever what they called the Great World War [6 thousand years ago] came along. Historians tell us, and the DNA confirms this group was also called the Arya and it is this group who invaded India according to the 'Rig Veda'[38]. They destroyed many, many of cities and pushed the Haplotypes identified as H and L [Dravidian Indians] downward towards the tip of India. This

---

*38 **Rig Veda**- This ancient history described what is commonly called the Aryan Invasion. The problem is the British arbitrarily placed a date of 1500BC on this segment of the Bharata War. Luckily, it has been more accurately timed to 5500 years ago by the shift in the Saraswati River in the texts.*

group of Scythians was identified as the R Haplotype who became R2.

For the Chaldean lineage of Nimrod, he married his Anakim mother named Semiramis. Here is a sample text.

Chronicles of Jerahmeel 33*- The beginning of Nimrod's reign was in Babylon, and there Nimrod became Bel. Cush begat Nimrod [around 7200 BC by Semiramis]. When Abraham [born around 2060 BC] was ten years of age, Nimrod's wife, Semiramis, reigned in Assyria forty-two years. After her there reigned Shim'i, who built the city of Babylon.*

By this we can believe Nimrod lived well over 5 thousand years and, while there is other documentation, we are told Semiramis was Nimrod's mom. Nimrod lived a very long time; and Semiramis ruled 42 years after he died and we also know Cush's Semiramis marriage was a disappointment to Noah, so we can believe she was not a normal Cro-Magnon.

In Egyptian history, King Thoth describes himself as living well over 10- thousand years and from the Island of Undal which was part of Atlantis. He goes into substantial detail how he as an Anakim human had a special method to live longer than Cro-Magnon. This strange method may have led to Thoth being hit by the horrible mutation during the Bharata War. As I went over briefly.

All the above discussions are saying this. Five thousand years ago, <u>something changed drastically</u>. Huge mutations showed up in the can Haplotypes. Somehow, America was settled <u>without going through Asia</u>, yet, here we are. My guess is air travel used to be very popular, but after the war, so many things were forgotten and the remains of any evidence has long since rusted away.

Lizard-Man Mutation

The last 4 mutations of the Bharata War period were the most devastating. The first was a lizard looking human some call the Ubaid because of the massive numbers of lizard-men effigies found there, but this mutation affected people as far north as England to the far East and throughout the Middle East.

Quit laughing right now are just skip to the next mutation without knowing how very common this mutation must have been. While 'Jasher' indicated some were turned into apes and elephants during the war. These reptilian looking people may have been what it was talking about concerning elephant-like. This type of human must have been mostly human with a very strange looking face. Coffee-bean eyes, tiny nose, pointed face and in many cases we find bumps on the arms. There is no question these people could still procreate and had children that looked like them. They were well received in society and there must have been thousands. Strangely and possibly luckily, they disappeared after about 500 to 1000 years. I don't want to get into these people in detail, as we don't have DNA to test, but here are some of the hundreds of effigies and images.

These unfortunate people showed up everywhere. Still completely human except for some strange facial and body features. Most were accepted into society.

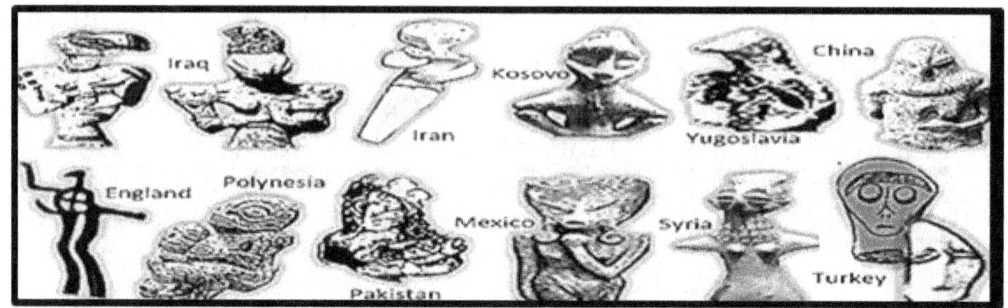

It seems like, for a time, just about every community had some of these reptilian-like survivors. They looked like they were from another planet. Usually, these people were well accepted into the communities, so we know the mutations were not excluded from the high levels of society.

A more devastating mutation turned man into ape.

Pan-Troglodyte Mutation

While it is OK to feel sad about these people remember we are also told that as many as 1/3 of the populations of the world would be killed during this horrible season of war and some would have an even worse mutation. That brings us to a DNA mutation called the Pan-Troglodytism. As I mentioned briefly, there were 2 really substantial major mutations. These were even more debilitating than the Ubaid lizard man mutation as they pushed back a human genome back to the Australopithecine type ape-man with almost no man. Two variants are still known including Chimpanzee and Bonobo. Like the Ubaid mutation these 2 occurred about 5500 years ago. The following graphic shows Australopithecus, Homo-Erectus, Cro-Magnon skeletons compared to the Bonobo and Chimpanzee and the skulls of bonobo, followed by a normal human, and a chimpanzee are also shown. The last 2 graphics show how these 2 apes are more human than gorilla in DNA. The last graph shows the 2 major types of human races, the Neanderthal, Denisovan and the Chimpanzee regarding major displacements of DNA structures. There are other books out about this particular tragedy.

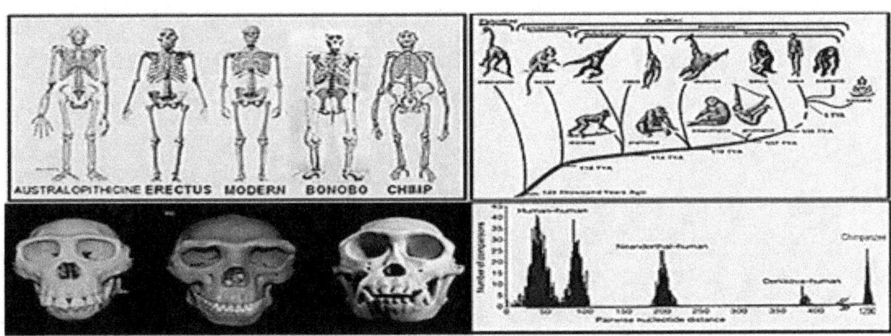

The feet had reverted back to hands like the Australopithecus, The arms were extended and the lags were shortened, the face became ape-like and the brain had been shrunk significantly. Possibly, the worst part of this mutation, is that it did not go away. We still had Chimpanzee and the Bonobos, but they are only found in a small section of Central Africa today. If you choose not to believe this, it's OK, as it doesn't change the details of this book. Just imagine looking at your new child and there was a monkey or even a lizard-like baby. Let me tell you there was sadness after the wars around the world.

Let me go over one more worrisome mutation of human DNA that may still be around today. The Indians called the mutated people Vanara and in Africa there were called Nomoli. Many were heroes and some were even kings, but they looked a little strange. Today they are classified as Homo-Cognatus and the mutation was common along the same latitude as the massive 500 square mile glass filled crater between Libya and Egypt. Children were born not with a lizard face or as an ape. Instead they became ape-men. These mutated apish-men were bright and courageous, but they certainly had issues.

Ape-Man De-evolution

Between 3200BC and 2000 BC something very strange happened. Ape-men appeared everywhere in the world and many became revered heroes, military and security leaders, and even kings. King Thoth of Egypt, for instance, had been "changed" to be ape-like, the King of Teotihuacan./Mexico Quetzalcoatl was changed to an Ape-man, General Hanuman of India, and King Sun Wukong of China are a few of the well-known rulers as shown next.

Let me talk about something that is very curious and related to this whole evolution and modification of Homo Sapiens and Erectus DNA subject. Humans mostly mutated into races just before the end of the Pleistocene and a second time around 5500 years ago during what was called the Bharata War during which the book of "Jasher" tells us 1/3 of the population of the entire world lost their lives and another third became like apes and elephants. Certainly, the Pan-Troglodytes were like apes, but there was an in-between as well. The Indian ancient histories called them the Vanara. I know this sounds stupid so let's look at this and other similar texts. While some of these could be representing the Pan-Troglodyte mutation, but remember that only was found in a tiny portion of Central Africa.

***Jasher 9:35-39** -those who said, we will ascend to heaven and serve our gods, <u>became like apes--</u>*

***Totonac- Mexican Tradition**-After the flood, the boat finally rested. Later <u>God reversed man's face and hind parts and turned him into a monkey</u>.* [This is probable an indication of man becoming ape-like during or after the war.]

***Mayan Tradition**-During the second creation, <u>people turned into monkeys</u> and the world was destroyed by wind.* [This is probable an indication of man becoming ape-like during or after the war.]

***Aztec History**-During the age of the four winds <u>men turned into monkeys</u> according to Codex "Laticano-Vatino"* [This is probable an indication of man became ape-like during or after the war.]

***Lower Congo Tradition**-"First God created man. Sometime after a huge flood, <u>men put their milk stick behind them and were turned into monkeys."</u>* [This is probable an indication of man turning to a chimpanzee during or after the war.]

***Tibetan History**-"Tibet was almost totally inundated by the flood. Later, the <u>survivors had been little better than monkeys</u>. The god Gya sent teachers to civilize the people and they repopulated the land after the flood."*

We are not sure what caused the DNA Mutation from man to ape-man, but we know it happened during the worst worldwide war of the Holocene Age, called the Tower of Babel War or the Bharata War. Whatever it was called, dozens of scientific studies all place it happening from about 3500 to 3100 BC. Effigies of ape-men are placed in farmland of Western Africa to show veneration. In Cambodia, they have half a dozen ape-men guards depicted around Anchor Watt. Multiple histories from India describe how entire villages of the ape-men became an Army that protected the Dravidians. In Egypt we find hundreds of depictions of these ape-men guarding the most scared items and being part of religious ceremonies. On and on we could go thought thousands of depictions of a horrible human mutation, just like the sacred book of Jasher claimed. It stated that *during the Tower of Babel War 1/3 of the population of the world died and another 1/3 became like Apes.* Here is a tiny sampling of the ape-men of America, Indonesia, West Africa, Egypt and India.

Scientists called this mutation Homo-Cognatus, while in India they were called Vanara, and in Africa they were called Nomoli. The following map shows the locations of the various Babel war mutations including the Homo-Cognatus mutations. Strangely, all are along the same latitude a massive 500 Square miles area in the Libyan Desert that has completely turned to glass. The graphic shows a tiny piece of the greenish glass and the map shows the huge area still covered with this melted sand.

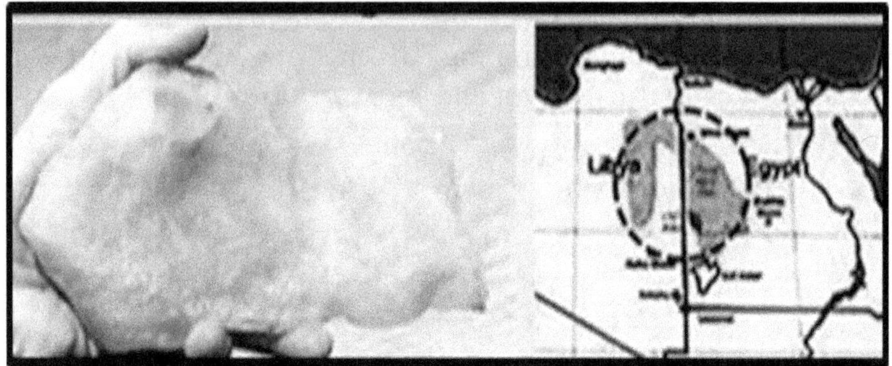

The crater has no indication of a meteor, so most believe it was some type of nuclear detonation. It would make sense that the radiation fallout from such an explosion, would mutate DNA. Please look at the general locations of the Homo0Cognatus mutations.

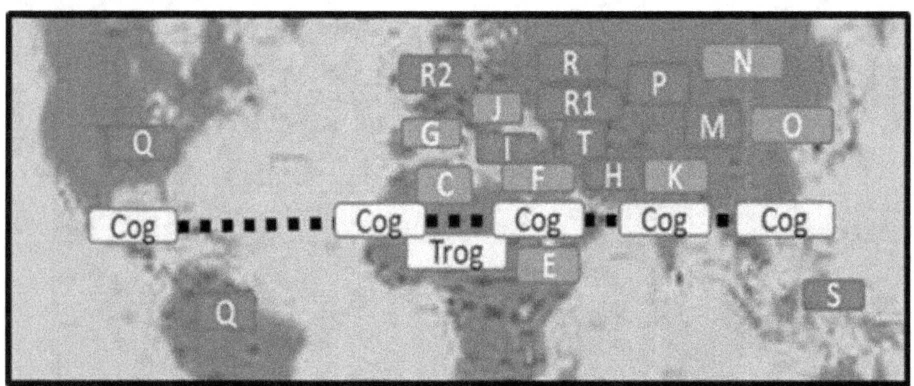

What if some of the Homo-Cognatus, those affected by the ape-men mutation survived to the present day?

Homo-Cognatus Today

Possibly when you saw the Vanara images, you thought of the strange sightings of ape-men of today. Most of the thousands of sightings of this human mutation known as Sasquatch, Homo-Cognatus, and so many other names, but a few are not only sightings, they included close inspection and display of their dead.

In 1870s, a sasquatch-like human mutation later named Zana[39,] was captured in Russia where she was forced to be a human display for over 20 years and had as many as 6 children by those who supplies this unfortunate, girl with whiskey. She was "very big and strong; could not speak; would not wear clothes of any kind; cold didn't bother her at all, as she was covered in Auburn hair, similar to the Paracas hair color and that of the Dolichocephalic Egyptian rulers; but she had negroid features and her offspring looked more African in features as they seemed to retain the human features of the rapist men. Her descendant's DNA and the DNA from a tooth from her son both concluded she had a substantial African Lineage They all had very wide eye sockets, an elevated brow ridge and what appears to be an additional bone at the back of the skull —one of her descendants was tested and showed she somehow was sub-Saharan. The following images include 2 depictions of Zana, one of her son followed by his skull, her granddaughter, and a newly viewed female-Sasquatch relative seen in the USA. About 1890, Zana died. She had been living in captivity, drunk and childless, for almost twenty years. Zana's unique features rapidly earned her celebrity status. No-one had seen anything like her. She was six-and-a-half feet tall, with black skin cloaked entirely in thick auburn fur. Zana also had more musculature than even the strongest men.

---

39 From "The Nature of the Beast"- The first Genetic Evidence on the survival of ape-men, Yeti, Bigfoot, and other mysterious creatures into modern times. Book by Geneticist/ Professor Bryan Sykes. 2015.

"

Her sheer strength was legendary, and she did not seem to be affected by the frigid winter air, even though forced to sleep in the outdoors.

In 1894, a Sasquatch was killed by trappers and photographed by Tom Biscardi, in Canada. The photos circulated for a long time when all but one disappeared. See image number one following.

In 1965, one Sasquatch human was found in Siberia and got to America, where it was exhibited by Frank Hansen, who went around the country. I was one of the lucky ones who was able to see this ape-man in 1968 identified as "The Siberskoye Creature". The next there images show this mutated man. It was also called the Minnesota Iceman and I saw it as a boy. Instead of being called Homo-Cognatus its formal name adopted in 1968, by a Belgium Scientific Journal, was Homo-Pongoides; but, it is now believed to have been very similar to Cognatus. Unfortunately, this example disappeared in the early 1970s. It is not known if the original is still extant or even if the model still exists, but as a boy I did get a chance to see this Vanara and it looked remarkable to me.

 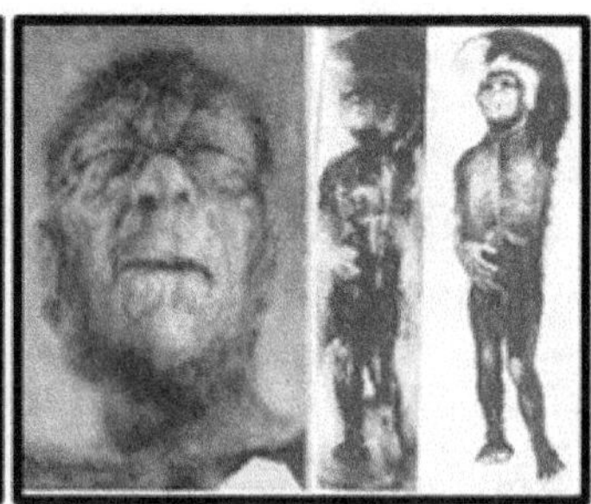 

In 2014, a Sasquatch tracker named Rick Dyer killed one of the ape-men in San Antonio, California. The corps was nearly 8-feet-tall and weighed more than 700 pounds. DNA testing on more than 100 hair and tissue samples showed this Bigfoot mutation was a close kin to human beings, according to Melba Ketchum of DNA Diagnostics. Last 2 images above.

More Sightings

I'm not getting into this section in any detail except for the DNA ramification as I haven't seen one of these unfortunate people and I know that trying to fool others is a favorite pass-time for some. There have been, however, well over 26 thousand sightings in almost every State of the USA, showing intelligent, upright walking, modern human hand and feet, ape-ish-men, somehow, were concentrated in North America even though evidence indicates, the major mutation of the original people was concentrated in the Africa and India. Hopefully, you have a better idea how the huge concentration of these individuals got to the Americas as Hanuman's army spread north and south from Patala Loka. There have been about 2-thousand sightings in Washington and California and well over a thousand from Pennsylvania, Michigan, New York, and Oregon with lesser amounts in the other States.

Besides hearsay evidence, there is a lot of video, footprint castings, and DNA evidence of what scientists are calling Homo Cognatus humans. Commonly called Sasquatch, Yeti, Yowie, Yeren, Didi, Masaabe, Seatco, Nunavut, we are talking about an 'ape-man mutation'. Thousands and thousands of sightings have yielded a substantial amount of hair, fecal matter, and all the rest so "its" existence could be tested. In 2012, Dr. Melba Ketchum's research in the Sasquatch Genome Project, included nuclear DNA studies of hair chromosomes. Her team sequenced 3 complete nuclear DNA genomes and 20 complete Mitochondrial

DNA samples. It was discovered that the nuclear DNA was very close to modern DNA, while the MtDNA showed about a 1% variance of unknown origin cross compared with 3 different samples in three different labs. Additionally, the genomes aligned with one another, meaning they came from the same species. The mitochondrial DNA of the specimens looks human, but the nuclear DNA (male lineage) was a structural mosaic consisting of both human DNA and novel non-human DNA, which <u>did not match any known species in GenBank</u>. The human-like DNA suggested this separation of species did not occur long ago; perhaps, 5500 years ago when Hanuman and some of his army traveled north and had children with the Adena. While sightings are much more frequent in North America, these people are sighted in forested areas around the globe as shown in the following tiny sampling.

Because the Nuclear DNA, generally, matches ours, the Homo-Cognatus is a blood relative. Also it was discovered that some unusual mutations are very different than Neanderthal or

Denisovan. Here are some blow up and enhancements of these apish people.

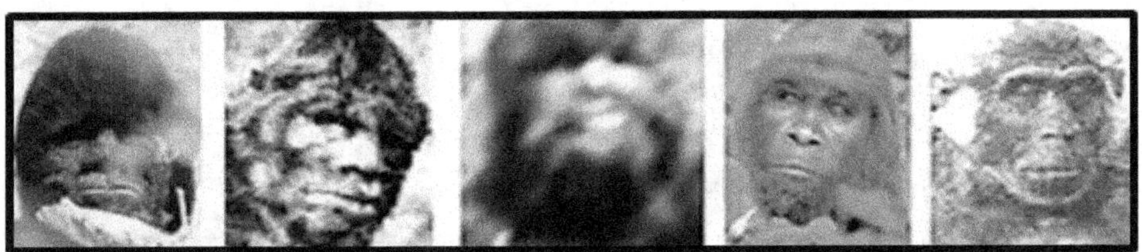

Most likely, this mutation happened the same time that the Troglodytes and the 12 new major human mutations occurred 5500 years ago. I think we are seeing today, the tiny fraction of the hybrid descendants of remaining Vanara/Nomoli individuals and it appears there is some association with a small group who came to Patala Loka.

Ape-man Hero of Egypt-Let me quickly tell you the story as it might involve the famous King Thoth of Egypt. He wrote about how he would have to be reincarnated every thousand years or so and he had a machine to help him do just that. His journal[40]. It seems that all of a sudden all images of Thoth were changed to an ape-man image as if something happened during the War and then he disappeared. Here are the progression of his images. The curved beak of the Ibis was supposed to show his association with the crescent moon. The ape-man has no reasoning.

---

40 Emerald Tablets- These make up sort of a very detailed diary of the first Egyptian King of the Holocene Age and the designer of the "Great Pyramid". In it he apparently is in his chamber simply called the Place of Reincarnation, when the War was underway

Ape-Man Hero of India-All of a sudden, a similar ape-man image with similar capabilities to that of Thoth appeared in India to save them from destruction. This ape-man was named General Hanuman.

Ape-Man Hero of Patala Loka-Now for the Indian Story as Hanuman left India towards the end of the war and traveled to the other side of the Earth to a place called Patala Loka and never returned. I won't get into details but there is quite a bit of evidence that Patala Loka was the City of Tenochtitlan in Mexico. All of a sudden the God-king of these PreMaya was depicted as an ape-man named Quetzalcoatl, The adventures of Thoth, Hanuman and Quetzalcoatl have a strange similarity and all this happened near the end of the worldwide war. Here are the images most often used for these men or man.

There is no doubt, from testimony and artifacts, that the major area of mutation was between India and Egypt, but some individuals went east to the Americas where the most ape-men sightings have been collected. I know all this ape-man stuff is still hard for you to believe, but just about every society is telling us that it did happen and anthropologists have known all these stories for years and never presented them. You have to wonder. All they did is say all the investigators, around the world, were crackpots and they know more, even though some of the historians may have had first-hand knowledge or heard the same stories over and over again. All I can say is there seem to be more sightings every year and I have to believe that since, Chimpanzees and Bonobo mutated from humans and a massive number of ape-men mutants roamed the earth and were generally accepted in society for a time, that some of them must be still out there.

As we leave this colorful section, let's look at a more detailed big picture of DNA haplographic tracking as we see the explosion of humans around the world and who they probably were related to in some way at three different times 20,000, 9000, and 5000 years ago.

That being said, there is still a lot we can learn about DNA and what it actually does besides being the simple instruction sheet for cells to use whenever they are trying to make a liver, or a spleen. As I mentioned before some of its capabilities has to do with Bio-photonic communication.

 Let me warn you that some of the next section will be difficult to understand as most, including me have difficulties. Even though there will be confusion, you will learn something about DNA.

What Do Bio-Photons Do?

Photons Control Everything in Living Organisms-It seems Photons switch on and control ALL the body's processes. Given different frequencies, identical cells perform different functions. The question that has forever puzzled cellular biologists for decades has been the following:

"What is it that enabled the tens of thousands of different kinds of molecules in the organism to recognize their specific targets?"

We now are beginning to understand how it is happening. It's not helping us define what photons are and what light is, but it is given us the details we need to begin to construct a definition.

Let's put it bluntly; Bio-Photons produce or establish life, but we don't really understand bio-photons. They explain how enzymes can recognize their respective substrates, how antibodies in the immune system can grab onto specific foreign invaders and disarm them. By extension, that's how proteins can 'dock' with different partner proteins, or latch onto specific nucleic acids to control gene expression, or assemble into ribosomes for translating proteins, or other multi-molecular complexes that modify the genetic messages in various ways.

It seems that, <u>somehow,</u> each molecule <u>sends out a unique electromagnetic emission</u> that can "sense" the field of the complimentary molecule. By this, molecules recognize their particular targets and vice versa by electromagnetic resonance. In other words, the molecules send out specific frequencies of photonic/electromagnetic waves which not only enable them to 'see' each other, but also to influence each other at a distance and become drawn to each other. With about 100,000 chemical reactions happening in every cell each second and each one initiated by some special coding of bio-photonic emissions, you can see that photonic energy is switching on and off continuously. One researcher put it this way---

It has been suggested that the way DNA is coiled is to allow it to change its "resonance" to send and receive various frequencies needed to support life. The image tries to show how the DNA can tighten it coil to vibrate more efficiently at a higher frequency or slow depending on internal and external requirements and you just thought DNA was a string of sugar.

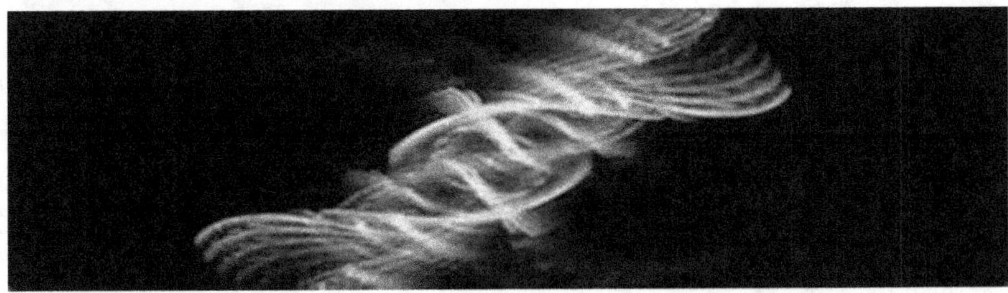

As shown previously, like a tiny spring, higher frequencies would have it tighten while low frequency emission needs would have it loosen its coils to allow better transmission and better resonance which would allow a particular frequency to affect it more or affect it more. <u>It is also known that as the DNA uncoils, the amount of light emitted is increased</u>. We now can sense what happens when you get a cut or scratch on your skin. The images following show the extra photonic activity in areas of distress on our skin and anywhere else on or in our bodies. This is not just light. It is a message carried in the light.

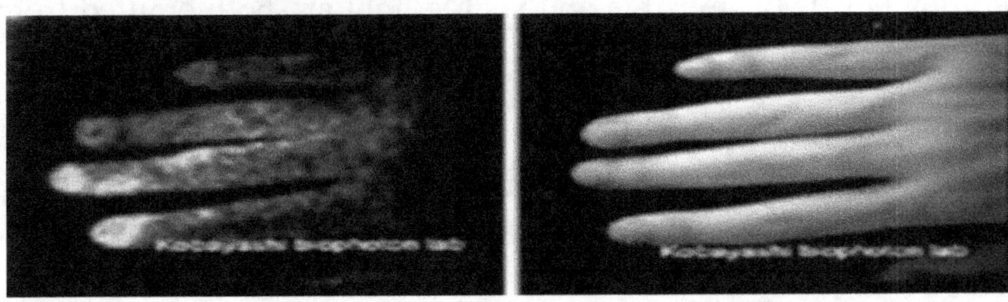

Please notice that while the last finger has distress, the adjacent finger is very active without the hand getting too much involved.

Now For the Halo

Sometimes the photonic emissions of DNA can be seen as visible light. If you remember from Religious histories, there was mention of Bioluminescence or HALOs. Those holding substantial amounts of "Faith" would begin to glow with something called a Halo. When Moses came down off the mountaintop after talking with God, he had one of these halos as bio-photons were emitted in huge amounts.

*Exodus 34:29-35-Now it was so, when Moses came down from Mount Sinai--that Moses did not know that the **skin of his face shone** while he talked with Him. So when Aaron and all the children of Israel saw Moses, behold, the **skin of his face shone**, and they were afraid to come near him. And when Moses had finished speaking with them, he put a veil on his face. But whenever Moses went in before the LORD to speak with Him, he would take the veil off until he came out; and he would come out and speak to the children of Israel whatever he had been commanded. And whenever the children of Israel saw the face of Moses, that the **skin of Moses' face shone**, then Moses would put the veil on his face again, until he went in to speak with Him.*

Today, scientists are finding out this is a truth as DNA when synchronized with hundreds of thousands of DNA in nearby cells can become very powerful. What needs to be considered is that very low levels of invisible and visible light are both emitted from the body and during times of stress, this output is increased. All organic life absorbs, emits, and processes light. Bio-photon emission or spontaneous ultra-weak light emission has been observed from almost all living organisms, with intensities ranging from 10^{-19} to 10^{-16}W/cm^2. A number of studies also found that emissions change by various cycles of the body. In Moses case, His head turned bluish. Possibly his whole body was emitting, but most of his body was already covered. Because he was wearing a robe, only his face was glowing, maybe his feet had a little glow, but the scary part was his face. Besides Moses, we find others had a halo of sorts and that was written down to show, our DNA can be activated and produce visible light at times.

Esdras Revelation[41]- *And on the throne sat one whose face was hidden. And there was <u>a rainbow around about the throne, which looked like emerald.</u> And round about the throne were thirteen seats: And upon the seats I saw thirteen elders sitting, <u>clothed in white raiment;</u> And <u>their faces were hidden by swirling clouds of light</u>*

Book of Dyzan[42] *- The first great waters came as stars showered down on the black-faced while they slept. The holy were the only saved. The unholy and most huge animals were destroyed. Few men remained. The king of <u>the Dazzling face</u>, was sad. He sent his air-vehicles, with pious men inside--Let <u>the dazzling faces</u> take the flying ships away from the lords of the dark skinned.*

Revelation of Moses[43]*-And Eve bore a son and <u>he was lustrous</u> and his name was called Cain.*

Genesis Apocryphon 6-*Then I considered whether the pregnancy of Noah was due to the Watchers or Nephilim ---he lifted his face, and <u>his eyes shone like the sun the face of this child was like fire.</u>*

The thing here is that DNA vibrates really fast to produce light. Usually this is ultraviolet light, but there are times, we can see the light-messaging that is produced by these very strange controllers of our bodies.

---

*41 **Esdras Revelation**- or Apocalypse of Ezra, is found in some Bibles and in the Essen collection of the Dead Sea Scrolls. It cross-compares with John' Revelation with a few added enhancements.*

*42**Book of Dyzan**- This ancient Tibetan works, mirrors the Genesis book with expanded details about the flood and similar world phenomenon.*

*43 **Revelation of Moses**- Also known as Life of Adam and Eve Apocalypse of Moses, 5 copies of this history was found among the Dead Sea scroll showing its importance.*

Bio-Photonics Has Taught Us

Before I get into this Star Trek stuff, let me just apologize, but DNA is more than just a pretty face. Let's take this whole DNA halo generator a little further. Now that physicist have added life-force of a cognitive-observer [Being with a soul] as a controlling dimensional element of our universe, we can see how DNA vibrational level affects the world in some way. We don't know the details, but there are all types of seemingly crazy examples showing how people levitate objects including themselves, lift cars off of people, sense remote sight while meditating, etc, etc, etc,. It is only know they are beginning to see how DNA is the controlling or focus element of all of these minor adjustments to reality.   I am not an expert in this area so I will leave it to physicists, but believe me when I tell you DNA and what it does is out of this world so to speak. Make us have a halo is child's play. If you remember in the Bible Jesus told his disciples they could change the position of mountains, turn water-in-wine, walk on water, heal sick people, and so much more with something he called "faith" and soon they were able to do some of that? All this is truly possible. It appears that, using quantum physics, this life-force control is REQUIRED to keep our reality –real.   And life-force control [let's call it regulation of our DNA resonance] not only affects our own bodies, but it "regulates" what reality is. As a life-force builds its resonant nature [making DNA vibrate faster] our "soul" focuses on controlling the environment more. Here are some bullets to think about.

- Light is associated with all life.
- Without light there is no life. Dead DNA cannot vibrate
- We can make a general statement that Cancer is too few coherent biophotons and Multiple Sclerosis is too many coherent biophotons, but both are controlled by

electromagnetic vibrations. IF we control that, we control these diseases.

- Stress Increases light output [perhaps as a plea!]. This is very common in plants.
- If plants were not stressed before being harvested, they do not output as much light. I don't know what that means, but it has been observed.
- If one does not have major issues in his body, cancer etc., light emissions in the tested areas has been seen to be low.
- Fluctuation in photon counts over the body is lower in the morning than in the afternoon, whatever that means.
- The upper extremities and the head region emit the most light and it increases over the day. I guess if the day lasted long enough we would all have a visible glow.
- Highest energy levels emitted by humans are 470-570 nm. This means that if one had enough people in a dark enough area and they were stressed, there would be a tiny blue haze that could be recognized. That being said, there are many wavelengths emitted by the cells as shown next.

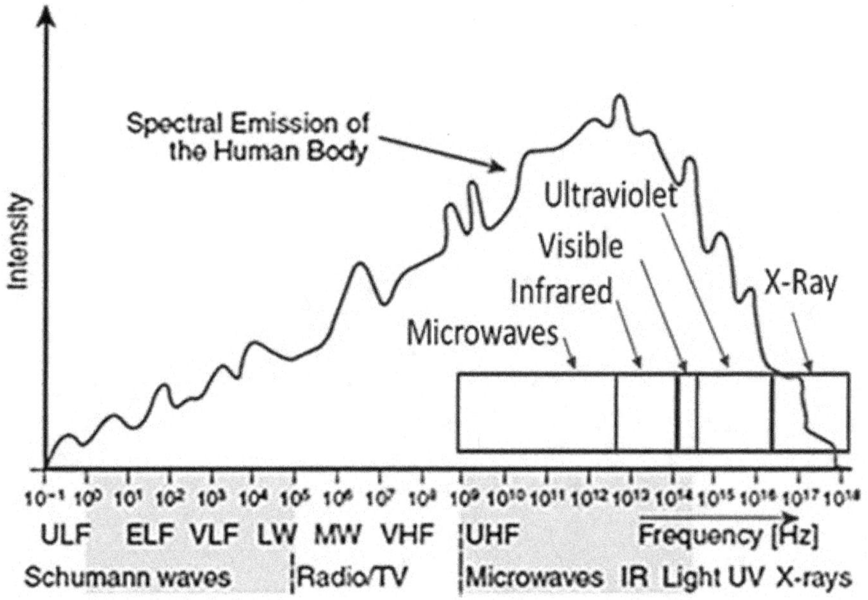

Photonic Emission Control

It has been suggested that biophotons connect us all to a sort of reality coordinator of some kind. This seems to really mean that everyone sends out all these messages so is a bunch of positive minded people are in a room, more positive things happen than if angry people are in an area. Scientists say it this way communal messaging drives our cognizant-anthropic or Quantum-fluctuating reality. I think it is something like this.

As we feel bad, that bad feeling modifies light emissions which make very slight changes to reality in a bad way, but if we force a positive attitude and get our DNA emitting better, Reality will be better. I've heard this called the Power of Positive Thinking. This is just saying having a positive focus when we are sick reduces how the sickness can affect us. Projecting that positivity is done the same way to others.

It has been known for some time now that our body, all other animals and all plants transmit and receive photons. Depending on how they are sent, things change in nearby cells or even nearby people or plants. We had better find out about this or we will never know what photons are. While the majority of the emissions are in the Ultraviolet range, some are visible. The levels are low as the photon communications are for short distances.

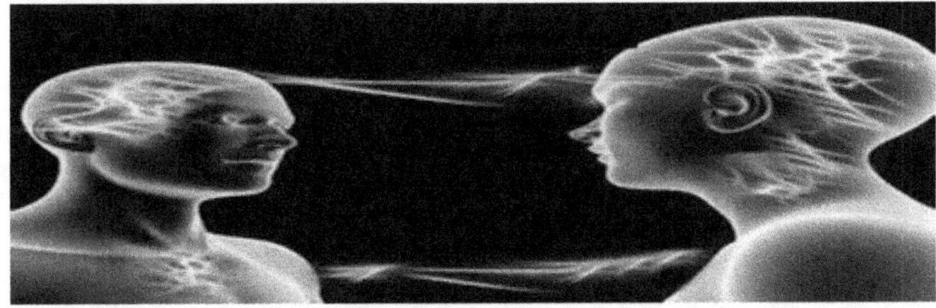

The preceding image might show how feelings and image transfers might look if we could see them. I think the best example is when someone cuts their hand open. The body needs to repair itself, so the affected cells send out a distress photonic message to the nearby "good" cells" that they need to replicate. The cells get that message, replicate, and, soon the cut is completely gone. The messages are by pulse coding and differences in output wavelength.

The most complicated and manipulative bio-photonic generator and receiver is in the molecules and known as DNA.

Photons Control Everything in Living Organisms-It seems Photons switch on and control ALL the body's processes. Given different frequencies, identical cells perform different functions. The question that has forever puzzled cellular biologists for decades has been, *"What is it that enabled the tens of thousands of different kinds of molecules in the organism to recognize their specific targets?"* We now are beginning to understand how it is happening. It's not helping us define what photons are and what light is, but it is given us the details we need to begin to construct a definition.

Winter Phenomenon-Everyone knows there is less light in the winter and more light in the summer. The question is, with less light, would there be more medical issues and Bio-phonons are not recharged as they would during summer months??? If we look at the chart below we can see that the answer is yes. Every winter, all sorts of disorders increase until the sun comes out again.

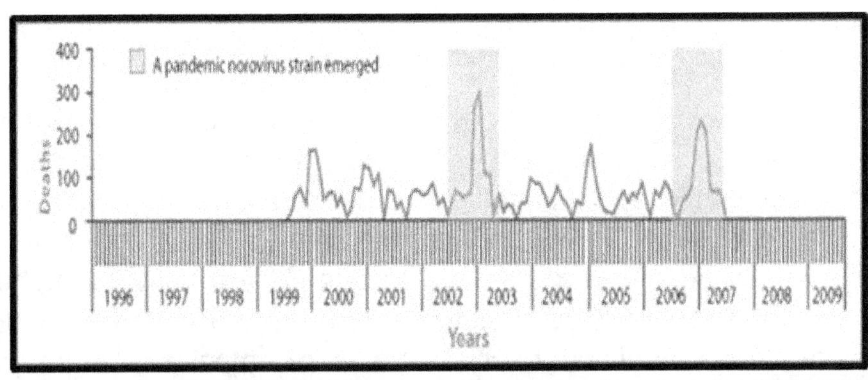

We get sicker in the winter because we don't absorb as much light.

American Journal of Preventive Medicine indicated the following: "*All mental health queries followed seasonal patterns with winter peaks and summer troughs.*" They found that mental health queries in general were 14% higher in the winter in the U.S. and 11% higher in the Australian winter.

Here is what researcher Dr. Fritz Popp had to say about this important topic.

"We now know, for example, that light can initiate, or arrest, reactions in the cells, and that genetic cellular damage can be virtually repaired, within hours, by faint beams of light. We are still on the threshold of fully understanding the complex relationship between light and life, but we can now say, emphatically that the function of our entire metabolism is dependent on light. This radiation from living cells, or bio-photon emission, represents a regulating energy field that encompasses the entire organism and affects the entire body's biochemical processes."

That being said the first attempt is something called color-puncture.

Color-Puncture and Skin Cures

Instead of sticking pins in your skin, why not just push photons into the skin cells. One attempt at using color light to save us is something called color-puncture. Similar to Accupuncture, this deals with photons. It is a holistic healing and one of Europe's most popular healing disciplines. The originator of Color-puncture is a German scientist, acupuncturist and naturopath named Peter Mandel who conducted over 25 years of intensive "empirical research" to develop this unique system of healing. Color-puncture involves focusing colored light on acupuncture points on the skin in order to energize powerful healing impulses in our physical and energy bodies. The images below show special feet, hands, and face color-puncture type devices.

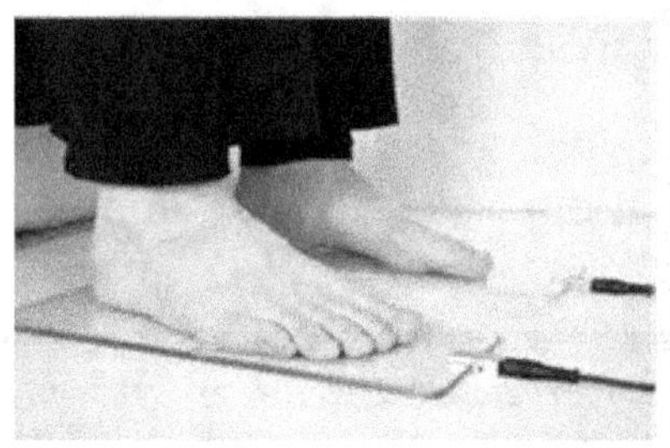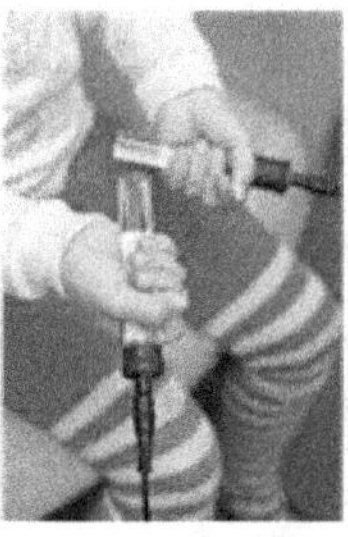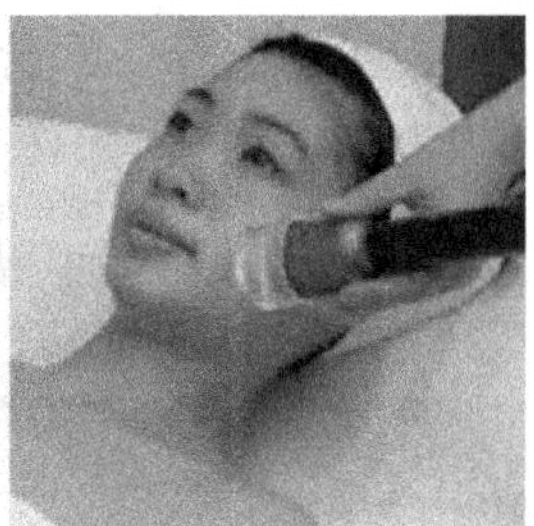

If you are really into it, more sestablished color theapists have massive colormachines to emit into your cells the life giving photons excite your DNA.

Curing Skin issues-Bio-photonic Light Therapy is on the rise and it seems to offer a vast quantity of benefits for skin DNA activation. There are a number of different color wavelengths that help the skin in a positive yet different ways. The layers of the skin are comprised of a high content of blood and water which allows them to easily and readily absorb and accept light. Once absorbed into the skin the different colors work jhard offering their ownunique benefits. Bio-photonic Light Therapy is safe and healthy and has no negative side effects.

- **Blue light** seems to reduce p-acne bacteria or excite DNA to accomplish this useful task. It can get through many skin layers, so there is little reddening of the skin.

- **Green light**, however, can go slightly deeper and helps in melanin production and reduce unwanted pigmentations by activing DNA in another way

- **Red lights** increases callagen production, and provides essential energy and DNA encouragement for celular renewal

The following images are for full body "help" as the main photonic emission of the body is red, most of these treatments seems to be with red light, but facial treatments are certainly looking into the blue and green levels.

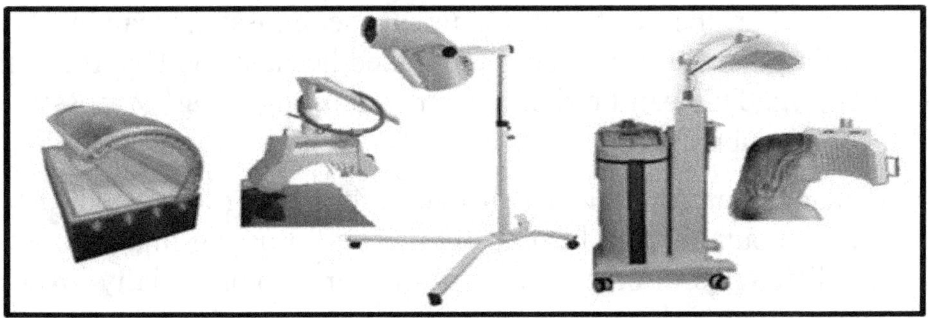

Patients report not only changes in their bodies, but improved emotional outlooks and a clearer sense of life direction after treatments. Patients usually report the following:
- Skin looking younger
- Fine lines gone
- Tightening of skin
- Less blemishes, Rosacea, and redness
- Faster burn scar reduction
- Less Eczema, Psoriasis

This gives us a good Segway into the next color treatment. For this we start with rocks.

Bio-Photonics in Plant DNA

Hopefully, you have started to understand the details presented as bio-photons can save your life, help you help others, allow for better understanding of people and plants, and possibly give us different direction for communication with others. So far, we have seen the following:

- Light is associated with all life. This includes bacteria, plants, and animals.

- Without light there is no life. Even after death there seems to be this outward indication of life for some time as bio-photons continue to emit by DNA.

- Even after the death of plants, these emissions can aid in fighting disease if they are not cooked because the DNA is still working. This can be done by injection, ingesting, or possibly even by rubbing.

- If plants were organically grown before being harvested, they seem to output more healing light. The following image shows the difference in organic mushrooms and commercially grown ones. To me it looks like organic food could help us more.

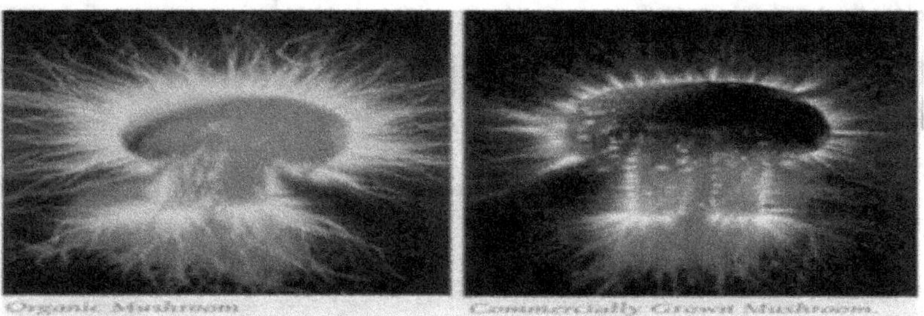

- We can make a general statement that Cancer is too few coherent biophotons and a huge bio-photonic output that is in some disorder. Cancer should be able to be treated by modifying the emissions of DNA.

- If one <u>does not have major issues</u> in his body like cancer, light emissions in the tested areas will be low. Cancer seems to increase bio-photonic emission of DNA in the stressed area so we can locate cancer easier.
- Everybody has a halo of very low intensity light emission from our DNA.

With all this emitting, do we need to turn ourselves off occasionally? The answer seems to make me yawn.

Someone found that <u>various vegetables would help flush the body</u> of these free radicals and reduce the possibility of DNA transmitting the wrong photonic message into the body. These vegetables may do even more. Let's take the old standard----<u>Mistletoe</u>. It may cure cancer.

Plant Cure of Cancer

Scientists around the globe studying the initiation and effect of bio-photons have begun to consider that your body's communication system might be a complex network of resonance and frequency. What if <u>vegetables actually changed the emissions of cells by combination and optical outputs like turning red light and green light into yellow light</u> or the stranger event of having red and blue photons combining to make <u>magenta?</u>

Magic Magenta

The only reason I brought up the last one is that **there is no magenta in nature** as blue and red are on separate sides of the rainbow. It is truly an oddball that has no wavelength on its own. What if adding fresh, <u>still radiating plants</u> into a person's body emitted signals that would combine with those aggressive outputs initiated by free radicals and the emissions together made a "color" that cancer cells simply could not understand? An advantage of this phenomenon is known as photo-repair. Let's see what happened with a <u>vegetable named mistletoe</u>. Here is what researcher Dr. Popp had to say again.

In one of numerous cases, Dr. Popp came across a woman in her thirties who had breast and vaginal cancer. He found mistletoe created something he called "coherence" in her cancer tissue samples. After a year, all her laboratory tests were virtually back to normal as the cancer had been "resocialized". Fresh mistletoe had transmitted photons into the body such that there was a direct or indirect modification of the cancer cell emissions.

Wouldn't it be nice to find a photonic message that could help circulation in diabetic patients? Maybe we can active DNA to establish better circulation.

Faster Diabetic Healing

This is very important Bio-photonic research focusing on increasing the life of diabetic patients by allowing their bodies to heal. Low-energy laser irradiance at certain wavelengths is able to stimulate the tissue bio-reaction and enhance the healing process. This is especially important when diabetes restricts new growth so drastically. Collagen deposition is one of the important aspects in healing process because it can also increase the strength of the skin. It seems that biphotonic irradiation increases collagen production needed for diabetic wounds in rats. The tensile strength of skin was employed as a parameter to describe the wound. The number of cases of diabetes mellitus (DM) worldwide is estimated to be around 150 million. This is predicted to *double by 2025* with the greatest number of cases in China and India. Diabetic foot and leg ulcer (DFU or DLU) is a serious complication of DM and is the single most important risk factor for lower limb amputations. More than 60% of all non-traumatic lower limb amputations are due to DFU complications. To make that even worse, around 50% of all non-traumatic amputations are as a result of DM. To further highlight the seriousness of diabetes associated lower-limb amputations, the 5-year mortality rate following amputation stands at 40 to 80%. Foot and leg ulcers are serious complications of Diabetes Mellitus (DM) and are known to be resistant to conventional treatment. They may herald severe complications if not treated wisely. That's where bio-photonics comes in. Electromagnetic radiations in the form of photons are delivered to the ulcers to stimulate healing. This first study was conducted to evaluate the efficacy of Low-Level-Laser-Therapy (LLLT) in diabetic ulcer healing dynamics. Once the photon energy is absorbed, the photo acceptor assumes an electronically excited state. One idea is that this stimulates cellular metabolism by activating or deactivating enzymes which can alter DNA and RNA. The energy which is absorbed by the photo acceptor can be transferred to other molecules giving us observable effects at a biological level. Photon energy is absorbed by cells, which activates secondary messengers and cascades the healing optical code.

Diabetic Rats

In one important experiment, diabetic wounded rats had large wounds induced by streptozotocin via intravenous injection. Skin-breaking strength was measured using an Instron tensile test machine. The experimental animals were treated with bio-photonic emissions at <u>808nm using a diode laser</u>. The photo-stimulation effect was revealed by accelerated healing process and enhanced tensile strength of wound. Laser photo-stimulation on tensile strength in diabetic wound suggests that such therapy facilitates collagen production in diabetic wound healing which will soon reduce suffering of diabetic patients.

LED Therapy

Besides these laser tests, lower intensity RED LED emissions have successfully been used to halt and reverse effects of this horrible life stressing condition. The following image shows one of the LED emission tools used for this type of Therapy

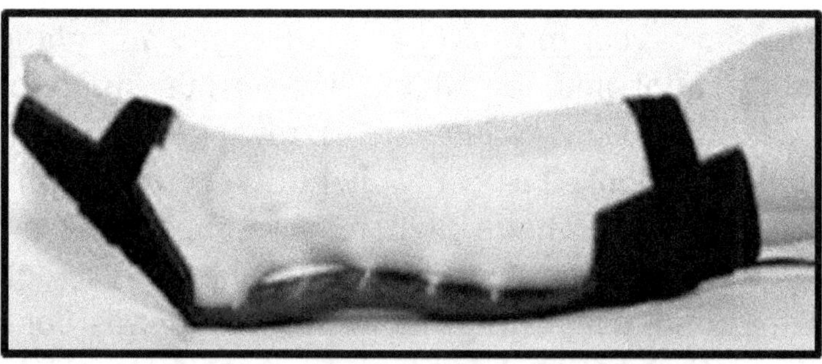

The diabetic ulsers on feet and legs are being removed by 670nm red light almost like the 808nm Laser light.

The next images show just how effective and rapid this type of theraphy is and no cutting off of legs and long hospital stays and loss of life savings. The firse image was of a diabetic patient almost ready for amputation as the leg was in terrible shape. The following shows a typical diabetic leg ulser treated over a 4 month time with "light". The ulcers almost seem to disappear before our very eyes as cells are messaged to change from a dying state to a rejouvenation state. Its like magic or laying on of hands.

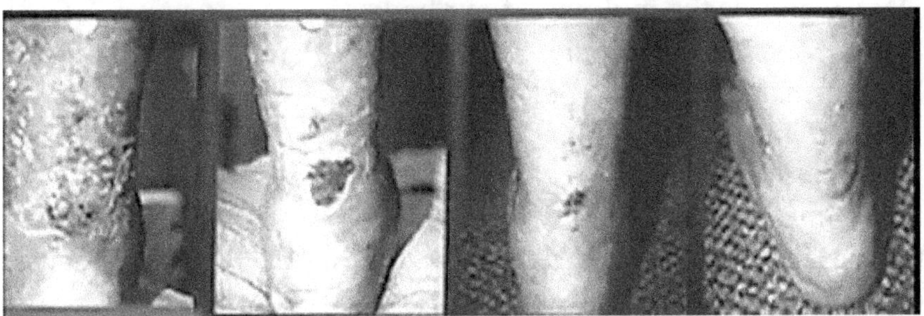

Sorry for the nasty pictures, but these are so much less nasty than many of the horrors associated with diabetes so get over complaining and understand how very wonderful DNA really is and how talking nicely to it can help us change our circumstances.

God and DNA

So here is what we have found so far.

- DNA Controls the building of our bodies
- It aids in sending chemical messaging around the body using hormones and Neuronic pulses in the brain
- We can help DNA in its messaging by sending external Phonic messages of the right wavelength.
- Sometimes DNA gets the wrong Optical messaging, especially from the sun that messes up the regeneration of new Cell. This can cause diabetes, cancer, and just about everything else.
- Even after dead, animal and plant DNA continue to try to fix themselves which shows up as a glow.
- DNA photo-messaging can change what stem cells will turn into or when it will change. [Salamander turning into a Frog]
- If we figure out how to change the natural resonance of our DNA photonic communications we can even make our skin glow.

Something called faith; not faith in our creator needed to go to Heaven, but a secondary type; seems very close to some type of amplified photonic messaging from DNA that we need to investigate. Before we end this topics, I want to expand your awareness of DNA to that associated with Biblical descriptions with the following particular aspects-as we are told the following:

- Faith of a grain of mustard-seed allows us to change our "reality" such that we can moved mountains. [Described by Jesus 3 different times]
- This same carnal faith turned water into wine or blood. [Jesus, Moses, Jane's, Jambres]
- Allowed a number of people to walk on water.[Elisha, Jesus, Peter]
- Allowed a number of people to heal the sick with their hands [Hindu monks, all the apostles, Jesus, Elijah, Elisha, Moses, several unbelieving Jews]

- Allowed a number of people to bring dead people back to life [Elijah, Elisha, Jesus, Paul, and Peter.]
- Allowed people to force out demons – [All the apostles, Jesus, several unbelieving Jews]
- Allowed people to survive a fiery furnace or boiling water. [The Apostle John, Abraham, and Daniel's three cousins
- Provides incredible Strength [A little girl lifts a car off her father, Sampson, and Shamgar]
- Running faster than a chariot [Elijah and Thomas]

On and on we could go, but does it have something to do with DNA? There are many references to people touching a sick person with their hands to cure them.

We also find 600 references to the Heart being the communication link between the Brain, Soul and Spirit in our complex {Cognizant Observer status among animals, so let's get a little more details about the hear

Thanks to recent scientific studies we now know the following:

- The Heart has its own brain. While smaller than the head brain, it operates and thinks 10 times as fast. I should be noted the Ant has a distributed brain similar to ours in that its antennae are filled with neurons.
- The heart brain communications extend ten times as far as those of the head brain. Therefore other people nearby can be affected.
- The Heart brain acts in a divergent way to head brain so it has cognition.
- The Heart, by controlling DNA emanations, seems to adjust hormone balance to calm or agitate the head brain.
- The three entities that make us up as a cognizant observer that can affect reality include the self, which has little effect, the Spirit which is almost exclusively an interface between this universe and the Heaven universe; and the soul. It is the soul that lives forever and established a link to what we call reality. It is the Heart that allows us to link up with the soul. Apparently this is done by optical DNA communications.

- Placing our hands on another allows direct DNA to DNA communication for healings of all sorts.
- Apparently, reality modification like changing the position of a mountain is not well understood, but the Heart DNA talks to the Soul which affects reality. Generally speaking, that is not enough.
- Apparently, the power of this soul-thing to affect reality depends on the body's resonant vibrational frequency to be as fast as possible. This "Resonant tuning requires as much of the bodies DNA to be working the same concern.

OK! I just put out some of the theoretical and observed elements of Heart and body DNA directed communication that seems to affect our very lives. If we want a better day, we need to believe we are having a better day. If we want to help someone else, we need to quit focusing on ourselves. If was want to move mountains--- I'm not sure, but Quantum Physics has already eliminated Time and space as constants, so I have to say while DNA and its communications is still a huge mystery, but there is no doubt that DNA can do much more than we currently can prove.

Conclusions

Hopefully you could tell that DNA is more special than a simple "Complex Molecule". One can say DNA builds the associated entity, keeps that entity working, and even after the entity is dead, at least in the case of plants. Even that type of definition doesn't come close to giving us insight into the important of these 2 meter long double helix string of sugar. Sorry for the esoteric items towards the end, but I think they are necessary for you to grasp the important of getting in touch with yourself to invigorate your DNA.

- DNA- Deoxyribonucleic acid is not really an acid. Instead it is s miracle of design, function, and life.
- The most interesting DNA are those associated with stem cells because a portion of the DNA is no programmed and can change into just about anything it wants to be.
- Some animals like salamanders have more Stem cell DNA than we do so they can regenerate their entire backbone, nervous system and even a portion of their brains. The problem is they don't have much to do with reality whether they live or die
- Our DNA have a far greater responsibility to our reality as we, somehow, talk to the soul portion of our being or allow our resonant vibrational frequency to be high enough to modify reality just a little. If our DNA is happy, we can bring happiness to others.
- Optical communication of DNA has been known to convert one species into another, change the growth rates of fetus, and all sorts of crazy things
- During the Pleistocene, the ancient people were doing just like we are today, but after thousands of years they got really good at it. The remade dinosaur and converted apes to ape-men
- Our creator created a man called Erectus and the Pleistocene people decided to change him to make him better.

- Finally God made a new man called Cro-Magnon. This man was special in that he not only combined updated characteristics, he also had a soul which allowed him to modify reality in a small way.
- The ancient people immediately tried to make this one better by spicing in their own DNA.
- War begat the destruction of the Pleistocene Age and more technological development afterwards ended up with the entire world in a season of war.
- The war was a disaster to mankind. They tried living underground, inside flying fortresses, in out space, and even under water. Like ants, they soon came to the surface, but now horrible disfigurement and changes in DNA was massive. A third of the people had lost their lives and another third ad become Lizard-like, Ape-like or a new form of ape.
- The mutation of DNA was so dramatic, lifetimes were shortened to less than 10 percent of the Pleistocene Ages, and brains quit functioning as well, so they atrophied. No-one could remember the once great societies and technologies and man was thrust back to a Bronze Age Existence. All this happed by mutation of DNA. This might have happened from nuclear radiation, or some germ designed in a lab, but it would be permanent.
- Climbing out of our hole, scientists focused on climate change without regarding DNA manipulation as the worst problem, they invented a new type of amphibian and turned a salamander into a frog. Behind the sense, experiments in Gene splicing made all type of weird animals.
- Every once in a while interesting things were found like how DNA used photonic communication. And a way to tell if someone was pregnant. Clones were constructed and human ears were made to grow on mice, but no one is seeing the bigger picture. Messing with DNA is not good for humanity.
- Don't get me wrong discovering that Stem-cells could and do turn into any type of cell they want which allows for reconstruction and all sorts of curative possibilities.
- Creating chickens with no feathers makes it easier to cook chicken and finding out how we can regrow backbone and the critical nerve tissues that could allow the lame to walk is interesting, but God told

everyone, if they would just have faith, they could do those things and much more by understanding our DNA rather than changing it.

Hopefully you enjoyed the book and possibly even learned a thing or two. Thanks for reading!

About the Author

Steve Preston is a long lime author of scientific and esoteric facts. His books focus on the painful truths rather than whitewashed details that make us comfortable. If you are interested in the truth instead of comfort, please review other works by Mr. Preston as shown in the following list. The images are some from Egypt taking the older version of taxi. To the right the writer is shown in the Jewish Negev desert of Israel where the Dead Sea Scrolls were found. I made these travels so you wouldn't have to.

To the left below are a couple of pictures as we searched the New Zealand caves possibly visited by the ancient Maori and the last image is of the author investigating the statues on the Acropolis in Athens Greece. I keep searching for the truth and when I find something was hidden, I try to get it to you.

History of Mankind Series	A New View of Modern History
20th Century To The End Of Time	Close Look at Ancient History
The Second Creation of Man	The Antediluvian War Years
The Creation of Adam and Eve	The First Creation of Man

Man After The Flood

Modern American Topics
History of Powerful Women
Promote the General Welfare
Modern Misconceptions
Our Very Odd Presidents
American School Disaster
The Bad Side of Lincoln
Can We Save America?
Great American Quiz
Humans on Display
Consensus Science
Monsters are Alive
US History Errors

Prehistoric America
Who Discovered the Americas?
Mysterious PreIncan Journey
Phoenicia and the Lost Jews
Romans found America
Crazy Quest of Ireland and America

Prehistoric Technology
Amazing Technology
Mysterious Pyramids
Incredible Titans
Anakim Gods
Not from Space

Prehistoric History
Creation and Death of Dinosaurs
Kingdoms Before the Flood
When Giants Ruled the Earth
PreBonze Age Kingdoms

Reality Science Anomaly
Our 12-Dimensional Universe
Mystery of Photons and Light
Meaning of Life and Light
Incredible Nikola Tesla
Is Time Travel Possible?

Biophotonics and Healing
Vibrational Matter
Slip Through a Wall
Anthropic Reality

Historical Fiction
Time Fixer the Beginning
Secrets of Washington
Time Fixer in India
Time Fixer in Egypt

Biblical History
Does Science Confirm the Bible?
History Confirmed By The Bible
Abraham to Moses
Adam to Abraham
Adam's First Wife
Moses to Jesus
Incarnations of God
Unbelievable Lineage of Jesus
100 Infamous Bible Women

Moses Studies
Moses Story Part 1
Moses Story Part 2
Expanded Genesis
Exploring Exodus
Exploring Genesis
Comparison of Genesis

Christian Studies
Understand the New Testament
Bible Revealed
Differences in the King James Bible
Why the King James Bible Failed
Understand the New Testament
Old Testament Used By Jesus
New Testament Mysteries
Errors in Understanding
Old Testament Mysteries
New Look at the Bible
Religious Anomalies

Funny and Frightening Bible Stories
Religious anomalies
Moses to Jesus

Secrets Series
Secrets of Daniel
Secret Life of Job
Secrets of the Disciples

Biologic Anomaly
Tracing Cro-Magnon to Jesus
God Didn't Make The Ape
DNA of Our Ancestors
Homo Erectus as a Man
DNA Anomalies
Races of Men
Lizard People
Secrets of Human DNA
Ancient Giant Humans
Creation and Death of Dinosaurs

Wars
America's Civil War Lie
Behind the Tower of Babel
World War with Heaven
Four Armageddons
World War Before
World War Zero
Six Deaths of Man
Driven Underground

Egyptian Studies
King Thoth
Truth About Hyksos Pharaohs
Scythians Conquered Ireland
Mysteries of the Exodus
Egyptian Foreigners
Moses Saved Egypt
Secrets of Thoth
King Joseph and the Secrets of Egypt

Metaphysic Science Anomalies
Beyond Virtual Reality
Releasing Your Consciousness
Understand your Heart
Vampires among Us
Awaken the Departed
Self, Soul, Spirit
Of Heaven and Hell
True Happiness
Photonics and Healing

Flight & Space Travel
Ancient History of Flying
Anomalies in Flight
Living on Venus
Space Anomalies
Where UFOs Go
Martians

Astronomical Anomaly
Return to the Moon Now

Modern Issues
Make your Own Global Warming
Terror of Global Warming
Fake Science
Behind the Death List
Can We save America?
Great American History Quiz
Bad Side of Lincoln
Humans on Display
Modern Misconception
Who Really Discovered America?
Truth About the Crusades
Allah' God of the Moon

Angels and Demons
Sex Crazed Angels
The Antichrist
The Devil